BEYOND CLIMATE BREAKDOWN

One Planet

Sikina Jinnah and Simon Nicholson, series editors

Peter Dauvergne, *AI in the Wild: Sustainability in the Age of Artificial Intelligence*

Vincent Ialenti, *Deep Time Reckoning: How Future Thinking Can Help Earth Now*

Maria Ivanova, *The Untold Story of the World's Leading Environmental Institution: UNEP at Fifty*

Peter Friederici, *Beyond Climate Breakdown: Envisioning New Stories of Radical Hope*

BEYOND CLIMATE BREAKDOWN

ENVISIONING NEW STORIES OF RADICAL HOPE

PETER FRIEDERICI

FOREWORD BY KATHLEEN DEAN MOORE

THE MIT PRESS CAMBRIDGE, MASSACHUSETTS LONDON, ENGLAND

The MIT Press would like to thank the anonymous peer reviewers who provided comments on drafts of this book. The generous work of academic experts is essential for establishing the authority and quality of our publications. We acknowledge with gratitude the contributions of these otherwise uncredited readers.

This book was set in Stone Serif and Avenir by Westchester Publishing Services.
Printed and bound in the United States of America.

Library of Congress Cataloging-in-Publication Data

Names: Friederici, Peter, 1963– author. | Moore, Kathleen Dean, writer of foreword.
Title: Beyond climate breakdown : envisioning new stories of radical hope / Peter Friederici ; foreword by Kathleen Dean Moore.
Description: Cambridge, Massachusetts : The MIT Press, [2022] | Series: One planet | Includes bibliographical references and index.
Identifiers: LCCN 2021049706 | ISBN 9780262543934 (paperback)
Subjects: LCSH: Communication in the environmental sciences. | Communication in climatology. | Climatic changes—Public opinion. | Narration (Rhetoric)—Political aspects. | Narration (Rhetoric)—Social aspects. | Denialism.
Classification: LCC GE25 .F75 2022 | DDC 304.2/50141—dc23/eng20220429
LC record available at https://lccn.loc.gov/2021049706

10 9 8 7 6 5 4 3 2 1

CONTENTS

SERIES FOREWORD

This is at once an odd and exhilarating time to be alive. Our species, *Homo sapiens*, has had roughly 350,000 years on the planet. For most of that time our ancestors barely registered as a quiet voice in a teeming chorus. No more. Now, a human cacophony threatens the ecological foundations upon which all life rests, even as technological wonders point the way toward accelerating expansion. We find ourselves at a moment of reckoning. The next handful of decades will determine whether humanity has the capacity, will, and wisdom to manufacture forms of collective life compatible with long-term ecological realities, or whether, instead, there is an expiration date on the grand human experiment.

The One Planet book series has been created to showcase insightful, hope-fueled accounts of the planetary condition and the social and political features upon which that condition now depends. Most environmental books are shackled by a pessimistic reading of the present moment or by academic conventions that stifle a writer's voice. We have asked One Planet authors to produce a different kind of scholarship. This series is designed to give established and emerging authors a chance to put their best, most astute ideas on display. These are works crafted to show a new path through the complex and overwhelming subject matters that characterize life on our New Earth.

The books in this series are not formulaic. Nor are they Pollyannaish. The hope we have asked for from our authors comes not from overly optimistic accounts of ways forward, but rather from hard-headed and clear-eyed accounts of the actions we need to take in the face of sometimes overwhelming odds. One Planet books are unified by deep scholarly engagement brought to life through vivid writing by authors freed to write from the heart.

Thanks to our friends at the MIT Press, especially to Beth Clevenger, for guiding the One Planet series into existence, and to the contributing authors for their extraordinary work. The authors, the Press, and we, the series editors, invite engagement. The best books do more than convey interesting ideas: they spark interesting conversations. Please write to us to let us know how you are using One Planet books or to tell us about the kinds of themes you would like to see the series address.

Finally, our thanks to you for picking up and diving into this book. We hope that you find it a useful addition to your own thinking about life on our One Planet.

—Sikina Jinnah and Simon Nicholson

FOREWORD

Kathleen Dean Moore

Human beings are the meaning-makers of the planet. This is our great glory. And this is our tragic flaw.

Standing at the edge of my porch, I can look over a long expanse of shoreline, a jumble of stones and rock wrack. Across the inlet, fog hides a heap of mountain peaks. Crows jostle at the lap of the water, cleaning a silver salmon's bones. What does it mean, this shining place?

I can imagine that this is a place of abundant and sustaining beings, each filled with life-urgency and beauty, each an essential part of a community of living beings at the fecund edge of the sea—a community to which humans profoundly belong. I can infer that my responsibility is to share the gifts, nurture them, use them wisely and sparingly, respect them and give thanks.

Although I struggle to do so, I can imagine, alternatively, that this shoreline is an inanimate storehouse of natural resources over which humans have ownership and control. No moral constraint stops us from selling the shellfish and salmon, mining the sand, and logging the forests until the place is empty and dead.

This matters—the meaning a culture makes of the world. Call it a worldview, call it a narrative or story, it answers the central questions of the human condition: What is the world? What is a human being?

How, then, should I act? The answers to these questions form a structure for our lives, dictating what is possible to see, what is possible to believe, even what is possible to imagine. The narrative may be a good, sustaining one. Or it may be an epoch-ending catastrophe. A culture will find out by living within that narrative and seeing where it takes it, in what we might imagine as an uncontrolled experiment on a global scale.

Here in the United States, the dominant culture is immersed in the metaphysical narratives that grew and evolved in Western Europe over centuries: first, that the world is inanimate, unfeeling, without value apart from its usefulness to human ends, dead stuff with monetary value only, most effectively put to use when transferred into private ownership by any means, including theft, force, or trickery. Second, that humans are the overlords of Earth—born greedy, competitive, individualistic—who are so technologically clever that they can destroy Earth without destroying themselves, creating short-term profit and foisting the costs off onto future generations. Third, that some people and cultures are of so little value that their lands and lifeways can be destroyed without loss, while other people, purportedly by virtue of their race or intelligence or modernity, are naturally superior and deserving of their privilege.

Few of us would acknowledge believing these stories, but few of us can avoid living in them; they have become as invisible and ubiquitous—and now as overheated and toxic—as the air we breathe.

In this important new book, *Beyond Climate Breakdown*, Peter Friederici argues that the planet has come to the end of a five-hundred-year experiment, testing what could be called the dominant worldview. The results are in. The conclusions are indisputable. The dominant worldview has led humankind to the brink of the irreparable breakdown of the planetwide systems that sustain human life and many other lives on the planet.

It is essential, he concludes, to recognize the assumptions that created the blueprint for the road to ruin and then to abandon them, to turn abruptly away before it is too late. It is essential to create new stories, or recover very old stories, that will allow us to live simple and joyful lives on a just and healthy planet. These are the stories that Peter Friederici urges us to create, in what must become the greatest and most urgent exercise of the human imagination the world has ever seen.

In the last ten years, the mottled starfish and sunstars have disappeared from this inlet. They have been taken by a virus made more virulent by the warming seas. And the water is indisputably warming. In the few decades since 1990, the ocean has absorbed an amount of climate change energy equivalent to three Nagasaki-sized atomic bombs exploding in the ocean *every second*. Across the Northwest, 95 percent of the historical populations of wild salmon and steelhead have been lost. The yellow cedars are dying, their roots frozen without the protective covering of snow.

We live in a time of dying. But the dominant worldview is also dying, challenged from every quarter by new narratives. There is hope to be found in the possibility of a sudden, irreversible transformation of perspective.

We are living through what philosophers call a paradigm shift, a time when the worldview that for centuries justified the power of the powerful, the cruelty of the cruel, the thefts of the thieves, the riches of the ultra-rich, is teetering and toppling. A time of paradigm shift is, of course, a time of bullies and shouting as powerful men claw after their vanishing privilege. This is promising: the louder the shouting, the closer we are getting to real change.

The failure of the dominant worldview may be terrifying, but it is necessary, because the empty space it leaves provides the opening for transformation. In this book, Peter Friederici calls on us to "step out of the yoke of the narratives whose comfortable weight we have allowed to settled on us over centuries and to find regenerative stories that link us to something new—and to something much older."

Once we are free of the destructive narratives, we will look back with weepy regret and astonishment at what the dominant worldview allowed us to believe. How could we, good and reasonable people, have been duped into believing the most self-destructive and world-destructive claims? In what possible moral universe are oil and gas companies justified in pursuing profit, even to the point of destroying the future? How does it make sense that industrial logging companies can clear-cut ancient forests and pour poisons on what remains, when the forests powerfully sequester carbon dioxide? How could it be right that people of color are forced to absorb the toxins spewed by the profligacy of the privileged? That people should be praised for amassing riches by selling or destroying what never belonged to them—the common heritage of clean air and water, a stable climate, ancient oil and gas and minerals, the extravagant beauty of the public land, the sweat of human bodies, the fertility of the soil, the extravagance of the oceans? That the government, established to secure the rights of people to life, liberty, and security, could be allowed shamelessly to collude with the powerful corporate forces that would violate those rights? How could we ever have thought this was right? How could we ever have believed this was reasonable? How could we have been so disastrously conned? The answer lies in the invisible network of false assumptions that has bound us for centuries.

"The ability to tell your own story, in words or images, is already a victory, already a revolt," Rebecca Solnit says. In the place of the narratives that forced us to think what now must seem unthinkable, *Beyond Climate Breakdown* asks for a wild multiplicity of narratives from imaginations set free: "At a time of crisis, undue adherence to present and readily understandable narratives is a sure path to ruin. . . . Our job is not to write an account of the future. It is to allow for the ongoing possibility that that story, or that web of stories, can be written later in an unending sequence that links past and future."

There is so much left to save. Earth is blessed with systems designed over millennia to persevere, communities of animate beings adapted to the race for life, and—most important—human beings endowed with the capacity to love Earth's astonishing lives and to imagine lifeways that protect their profound resilience. This month, the falling tide in the inlet has revealed something new: leather sea stars in great abundance. They look like bulging red stars and smell like garlic. In the absence of the great schools of herring, the humpback whales are feeding in a way people have never seen before. In shallow water, they roll on their sides, open their mouths, and engulf schools of perch or krill. Heavy with eggs, salmon are finding new rivers emerging from under melting glaciers. The question of unimaginable importance is whether—and how quickly—the human imagination also can change.

—Chichagof Island, Alaska

ACKNOWLEDGMENTS

A book that calls for addressing the prospect of climate breakdown with many-voiced stories can only have been birthed by a community, or numerous communities. It stems from the convergence of a number of strands that have made up my professional and more-than-professional life, and I am grateful to many friends and colleagues who have shared thoughts, provided insights, and in countless ways informed this project.

My writing grew directly out of classes I've taught at Northern Arizona University. The First-Year Seminar program allowed me to create a new class on risk, enabling me both to explore the literature on risk perception and to learn from my students how those sometimes abstract ideas play out in the real world, at least among eighteen-year-olds. The School of Communication supported my proposals for teaching science communication and developing not only a seminar but a graduate certificate centered on the practice of this fast-growing field. Sandra Lubarsky of the Sustainable Communities Program graciously invited me, years ago, to teach a pair of classes on ecological oral histories. For many years the program has also provided me with opportunities to learn from students whose passion for workable solutions to problems of sustainability and social justice never fails to inspire. I'm grateful to all these students, whose questions, suggestions, objections, and enthusiasm are thoroughly embedded in the arguments I make here.

Many colleagues, too, have helped along their way, both through concrete suggestions about the ideas presented here and through their own work on climate change. I particularly want to single out friends in the Sustainable Communities Program, whose consistent focus on the connections between environmental degradation and social justice has helped center my account: Kim Curtis, Nora Timmerman, Leah Mundell, Sean Parson, Luis Fernandez, and Janine Schipper. Diana Stuart and Brian Petersen deserve special mention not only for being remarkably productive researchers and writers with their own focus on similar topics but also for reviewing sections of the manuscript.

My partners at *Carbon Copy*, Lawrence Lenhart and Ted Martinez, gave me an opportunity to engage in a practical exercise in polyphonic climate change storytelling; I'm particularly grateful to Ted for his invitation to coteach an honors class on climate fiction and for suggesting the Invisible Man analogy and reviewing a draft chapter. Others at the university have helped create a fertile seedbed for exploring how climate change action can be put into place both at a university and in communities beyond its borders. At the Center for Ecosystem Science and Society, Bruce Hungate, Kate Petersen, and George Koch have created the space for discussions about climate science and community decisions. Darrell Kaufman, Dan Boone, and Victor Leshyk provided me with an early opportunity to share climate science through a film project on paleoclimate research. At the Environmental Caucus, Erik Nielsen, Caitlyn Burford, Stefan Sommer, Matt Muchna, Blase Scarnati, Avi Henn, and others worked to make connections between what we know and how we should act. Björn Krondorfer and Gioia Woods have helped fertilize creative and scholarly discourse throughout the university. Seminar invitations from the School of Earth and Sustainability, the College of Social and Behavioral Sciences, and the School of Forestry gave me what every journalist needs to finally put ideas together, namely, pressing deadlines.

Off-campus, editors and other writers have provided me with support, column inches, and encouragement. John Mecklin and Daniel Drollette at the *Bulletin of the Atomic Scientists* prodded me to report on Germany's *Energiewende* and came up with a witty headline that I've borrowed. Andrew Wisniewski and MacKenzie Chase, along with my comrades at the "Letters from Home" column, allowed free-range inquiries into place that came to be integrated into pretty much every chapter. Mark Rozema organized a "NonfictionNOW" conference panel that allowed fertile discussions of metaphor and science writing. Kathryn Newfont and Debbie Lee encouraged explorations of the connections between oral history and place. Deb Anderson provided encouragement on a chapter draft. Mike Crimmins helped me better understand the North American monsoon. Wolfgang Knorr and Rupert Read offered encouragement and stimulating ideas about the conjunction of climate realism and hope. A number of anonymous peer reviewers offered constructive critiques and suggestions at both the proposal and manuscript stages. Several editors at the MIT Press helped streamline both ideas and phrases. Andrea Ross and Annie Ross read the completed manuscript and offered helpful suggestions. Michele James crafted the figures. I also want to thank Peter Goin: by collaborating on climate change storytelling with a gifted visual artist, I'd like to think I learned to see with new eyes, at least a bit.

My exploration of the *Energiewende* and of how the media portray climate change were jump-started by the hospitality of Mike Schäfer and the School of Integrated Climate System Sciences at the University of Hamburg. Jan Linehan and Peter Lawrence at the University of Tasmania invited me to share my ideas in written and spoken form, as did David Holmes and Stephanie Hall at Monash University. I'm grateful to the NAU College of Social and Behavioral Sciences for the sabbatical support that made these enriching opportunities possible. My experience of drafting the manuscript in these recent months would have been poorer were it not for the "Existential Toolkit for Climate Educators" workshop sponsored by the Rachel Carson Center. Even though we didn't get to meet in person, the regular online meetings and information sharing inaugurated by organizers Jennifer Atkinson, Elin Kelsey, and Sarah Jaquette Ray have guided me to some helpful resources and been a regular source of encouragement.

My writing has also been aided immeasurably by those who have provided me with quiet places in which to write. I'm thinking in particular of Janine and Laurie in Hobart, Catherine and the gang at the Black Range Lodge, and Mark and Jill in Waldport, along with Sophie the dog. Those places are all embedded in the book. So are the many trails that are more important to me than desk space as I develop ideas. Though I often hike alone, I'm grateful to Patrick Pynes, Tim Bower, Chuck Larue, and my Flagstaff hiking buddies for their fellowship and ideas on many walks that helped shape my ideas. It is not too much, I think, to also credit the mountain and canyon trails themselves for speaking to me as I have worked out various connections. I couldn't have done it without them.

Nor could I have completed this work without the support and love of Michele and Liam, for whose encouragement and tolerance of the uncounted hours I've poured into this project I am ever grateful.

INTRODUCTION

"How dare you?" The question was the single most memorable line from teenage climate activist Greta Thunberg's testimony at the United Nations in September 2019. How dare the world leaders gathered there spend their time debating modest climate policies wholly incommensurate with halting the warming of the globe—fiddling around the edges, so to speak, while the world burned? How dare the world's governments have ignored or suppressed the ever-clearer findings of the scientific community for multiple decades? How dare the world's adults have dithered so long about climate change that they drastically reduced Earth's livability for their own grandchildren? And how dare they act so blithely about this act of betrayal through which they have imposed not only the physical consequences of their inaction but also the inevitable accompanying burdens of stress and anxiety on younger generations?

It was, and is, a great question, perhaps the principal question that has needed to be asked of society since the threat of climate change became obvious in the late twentieth century. In its close kinship with a more detailed corollary question—namely, what is the matter with a society that would willingly destroy its own future in this way?—it gets at the heart of this book.

It does so because the question invites narratives, because any possible true answer to her searching question has to come in the form of a story, and because story lies at the heart of our miscomprehension of climate change. We have not been able to come anywhere near dealing with it appropriately because a full understanding of climate change as an existential threat is too often swamped by myriad other narratives that distract us. As a result, it is not only the reliable predictability of the old climate system that we see breaking down all around us. So are many attributes of our social-political systems, including confidence in democratic decision-making and in planning for the future. And so are our cultural tools for assigning value.

It's true that an intellectual understanding of climate change as a genuine physical phenomenon that is altering our world today has become more widespread, even if achingly slowly in the case of the general population in the United States, and agonizingly slowly in the case of its most powerful lawmakers. One can only be encouraged by the degree to which young people worldwide are taking on this crisis as an organizing principle of their public participation and of their life trajectories. Yet the most powerful societies in the world have not managed to find shared narratives that would productively address our predicament.

According to the Yale Program on Climate Change Communication, in 2021, 55 percent of Americans were either "alarmed" by or "concerned" about climate change, leaving another 45 percent to fall into categories ranging from "cautious" to "dismissive."[1] In the political realm, to have more than half the populace concerned is often a sufficient spur to meaningful action. But in the case of climate change it has not been enough. In another Yale study, this one international in scope, large majorities in all the nations surveyed said they believed climate change is happening. But that belief doesn't always equate to support for action. In the United States, only 43 percent of respondents would contemplate participating in citizen action against climate change. This figure includes some of those who were "alarmed" or "concerned."[2] Meanwhile, the International Energy Agency predicts that without sharp political and economic changes, global carbon dioxide emissions are on track to reach a new record level in 2023.[3]

These reports are indicative of a confusion not only about what the problem is but, more profoundly, about how to respond. Many who advocate for reductions in fossil fuel emissions sufficient to meet what scientists say is required remain outside the political mainstream. Others, fearful of the political and economic ramifications of what needs to change, pull their punches. A vast middle ground is occupied by people who are concerned but confused, who don't know what role they can possibly play beyond making guilt-ridden changes to their personal behavior. Those who continue to deny the full magnitude of the problem, on the other hand, have no difficulty crafting increasingly implausible but still coherent stories explaining their position. As a result, a minority of the population that will not, or cannot, accept the full reality of climate change—or does, but hides that knowledge under a cloak of self-interest and nihilistic expediency—retains outsized influence. The rest of us are subject to all the pathologies inherent in lacking a coherent life narrative—anxiety, distraction, apathy, futility, withdrawal. Those responses are significant, and telling, for they demonstrate the core importance of story to our comprehension of the world and of our place in it.

I need to clarify that I am not referring here to the narrower concept generally dubbed "communication," vital though that is. In recent years scholars and practitioners have published some excellent guides to climate change communication, typically presenting practical ideas on how certain communication practices can overcome hurdles that stand in the way of understanding and acting.[4] This is not such a book. Rather, I am starting with a much broader conception, using the word (and the concept) *story* to refer not only to the myriad ways in which we as language-using animals communicate information, emotion, and values to one another but also as shorthand for how we aggregate the myriad sensory, cognitive, analytical, and linguistic inputs that we use to make sense of the world—a personal and even intimate process that doesn't end until we do.

The most important stories we craft are those we develop for ourselves so that we can make sense of the world and of our place in it, and this process is at work within both individuals and societies. I hold that this creation of *narrative*—a word I will be using to refer to a type of story with a particular explanatory purpose—is a culturally determined practice that we learn starting at a very young age from repeated exposure to how others frame their own stories. In media-saturated societies, it is extremely difficult to resist the influence and inertia of societal narratives so broad that they have taken on mythic status. And the framing practices we develop are so potent that they determine not only how we *explain* the world but also how we *experience* it, guiding us in sorting the incredible array of information coming at us in the course of a day, a week, a year, or a lifetime into patterns that we believe make some sort of sense. Climate change is a matter of patterns. One of the reasons we have failed to address it effectively, I believe, is that the narrative templates most of us in the developed world have grown up with and into, and which we regard as fundamental to our lives, are out of step with the patterns climate change is exhibiting. Yet even as the failed narratives we tell ourselves about climate have broken down, sticking with them has generally proved easier than undertaking the challenging work of developing new ones. As David Wallace-Wells put it in his controversial book *The Uninhabitable Earth*, "Almost everything about our broader narrative culture suggests that climate change is a major mismatch of a subject for all the tools we have at hand."[5]

The result is what might be considered humanity's greatest condition of gridlock. Even as it has become abundantly obvious that massive restructurings of our infrastructure, economies, and societies are needed to deal with the climate crisis, it has become equally apparent that our existing political and societal structures are not up to the task, whether in terms of practical results or in terms of the ability to imagine alternatives. In fact, it's unclear whether our prevailing cultural strategies for assigning meaning are capable of stretching sufficiently so that we can live a truly sustainable lifestyle on this planet. "The collective apprehension of the earth has begun to stop us from believing in any coherent future," the novelist Jonathan Lethem has suggested. "And if there is an adequate politics to that collective understanding, it certainly hasn't been invented yet, and it isn't even being attempted."[6] If our species were a person, it would be diagnosed as suffering from cognitive dissonance, unable to reconcile the disparate elements of its existence and looking for any means of smoothing out this painful schizophrenia through ever more irrational workarounds.

I hope it is clear here that I am writing about much more than outright climate change denial, which in its baldest form is confined to a relatively small percentage of the population—though that small percentage, unfortunately, is greatly overrepresented among the political and corporate elite, especially in the United States. Rather, I am concerned with a much broader variety of forms of denial or distraction, all of them potential barriers to action, from religious fanaticism to blinkered techno-optimism,

Romantic fatalism to paralyzing despair. Climate change is a disruptor of much more than the physical patterns of how energy moves in the biosphere. It threatens worldviews, and not only those of the people who deny it most fervently. "It would be a mistake to assume that denialism within the Anglosphere is only a function of money and manipulation," the novelist Amitav Ghosh writes in his book-length essay, *The Great Derangement: Climate Change and the Unthinkable*, which wrestles with the question of why it is powerful English-speaking countries that have had the greatest difficulty in accepting the reality of climate change. "There is an excess to denialist attitudes that suggests that the climate crisis threatens to unravel something deeper, without which large numbers of people would be at a loss to find meaning in their history and indeed their existence in the world."[7]

This essay is an effort to track down what that "something deeper" is. It is intended as an inquiry into how our use of narrative, following the well-worn ruts of habit, falls short in coming to terms with a threat the likes of which we've never faced before. It explores how our predisposition to certain forms of stories shapes what our senses perceive, and how the massive wave of information about climate change that has washed throughout our society has only slightly shifted public policy. Above all, it examines how language, one of the supreme achievements of our culture, has so far proved dangerously inadequate to describing where we are, how we got here, and where we need to go.

My work is in no way a denigration of any of the exciting and vital array of other work going on to address climate change—the reworking of environmental and economic policies, the development of new technologies, the massive mobilization of education and protest. That all needs to happen, everywhere, all at once. But those efforts will not succeed without a simultaneous effort to figure out how to talk about the problem and possible responses—or how to "apprehend" it, as Rob Nixon writes in his book *Slow Violence*, using a term that he says in its linkage of the meanings *arrest* and *mitigate* "draws together the domains of perception, emotion, and action."[8] Only by apprehending climate change in a way that links those domains will we be able to organize appropriate responses.

Before I provide an overview of the structure of this book, I want to note some important points about its foundations and language.

First, even though the book dedicates a fair amount of space to parsing interlinked forms of climate change denial, it does not in detail lay out the scientific case for why and how climate change is happening. This ground has been thoroughly trodden, and there are numerous sources available in print and online that explain why and how climate changes, how researchers measure those changes, and how they project possible futures. In my analysis I rely on the findings of the worldwide scientific community as expressed in reports of the Intergovernmental Panel on Climate Change, in numerous peer-reviewed research reports, and in popular distillations of that science that have appeared in books and periodicals. The counterarguments to this wave of science are interesting as story, which is why I examine them, but as science they have about as much validity as claims that smoking is not bad for health, or that Earth is flat. They rely on such stretches of logic, and such cherry-picking of evidence, that they are better regarded as attempts at crafting a counterfactual fictional worldview than as genuine efforts to help us understand what is actually happening in the physical world. I'm going to use the broadly shared basis of agreed-upon science as the foundation of my argument for why we urgently need to change our approach.

That basis includes as core elements the certainty that the greenhouse gas emissions we have long pumped into the atmosphere have already substantially changed patterns of temperature and precipitation; that those alterations have caused numerous spillover effects, such as rising sea levels and ocean acidification, drought, increased wildfire severity, and more extreme weather; that a great deal of future heating is already locked into the Earth system because of the amount of energy the oceans have absorbed, the time it takes for greenhouse gases to be removed from the atmosphere, and the inertia built into our energy infrastructure; that the warming that has already taken place, plus that sure to come in the next few decades, is likely to unleash physical feedback loops that will further worsen the problem; and that the coming levels of heating and ecological turmoil are very likely to cause enormous geopolitical convulsions. These aren't speculative notions. They are sober and thoroughly vetted analyses that have resulted in conclusions like this one, from a 2020 report: "Life as we know it is threatened. . . . Something will have to change at some point if the human race is to survive."[9] And that analysis was written not by an environmentalist or even by an academic scientist but by economists at one of the world's largest investment banks, J. P. Morgan.

Second, I need to address a stylistic usage I have adopted that some readers will no doubt find problematic—specifically, the use of first-person plural pronouns to refer to what I believe are widespread patterns of perception, analysis, belief, and action (or inaction). Though I sometimes use the word "we" to refer to the entire human species, I often use it—or "our society"—to represent the dominant societal thinking that prevails in the global west, and particularly in the United States. I use "our society" to mean, in the United States, the political power represented by the mainstream of the Republican and Democratic Parties; the neoliberal consumer and media society that most Americans, and a growing number of people in other countries, have bought into whether they like it or not; and the generally rigid vision of national interests that largely prevails at gatherings like the UN Climate Summit.

My usage reflects the privileged perspective of those who, thanks to geography, skin color, gender, or coincidence of birth, are in positions closer than many to economic, cultural, and political power—and further removed than many, at least so far, from the most immediate consequences of climate change. In my case, this doesn't mean that candidates I vote for necessarily win; they often don't. But it does mean that as a white male US citizen born into a comfortable middle-class background in the 1960s, I have with what seems to me something like ease been able to surf the ascending economic and political might of a particular neoliberal means of organizing society that has prevailed in the global west since the end of World War II, and across a much wider swath of the Earth since the end of the Cold War.[10]

But in parts of this book I unpack this "we" to show that it is a fraught amalgam of differing and at times contradictory values and viewpoints. Much of the power of Greta Thunberg's "how dare you" phrasing, after all, stems from the unsparing honesty of its you-versus-us framing, exposing as it does the profound divide between powerful older generations that have largely benefited from policies promoting climate degradation and the so far less powerful younger generations that are increasingly paying the price for their elders' profligacy.[11] My use of the words "we" or "us," or the phrase "our society," is not intended to elide these differences or to imply that minority social groupings are not important—they are critical, as I discuss more than once. Rather, it is intended to act as shorthand for political, economic, and cultural power and as a reminder that the great majority of people alive today in developed nations, and certainly the more privileged among us, are strongly tied in to the cultural, economic, political, and infrastructural means of being in the world that have led us into the climate crisis. It will be up to individual readers to decide how much, and where, they identify with or feel in opposition to my particular use of these collective first-person pronouns.

What "we" represents also has a lot to do with what sort of policies will arise in the face of climate breakdown. In the United States, the federal government's decades-long failure to act is inextricably linked to broad failings of our democracy, especially to such phenomena as the overinfluence of moneyed interests on voting and political decision-making and the underinfluence of those most immediately affected by climate change; it is linked as well to such structural factors as a political party system that rewards extremism and the persistence of the undemocratic Electoral College. I believe that meaningful responses to the climate crisis can succeed only if the words "we," "us," and "our" are made whole. They need to be broadened so that governments at various levels in the United States and beyond can for the first time truly claim to speak with the "consent of the governed," as Thomas Jefferson phrased it in the Declaration of Independence. They need to be broadened so that they include not just certain segments of human society but *all* human society. As Timothy Garton Ash asked about the future of democracy in a 2004 book, "What's the widest political community of which you spontaneously say 'we' or 'us'? In our answer to that question lies the key to our future."[12]

Just as important, the referents of those pronouns need to be broadened to include the other living organisms with which we share the Earth, for resisting the breakdown of the global climate is ultimately a test of the affinity that human beings have for life itself. It is the ultimate collective action problem that our species has had to face, and it demands an unprecedented degree of collective global identity—of agreeing that in acting to protect the future of life, we are all "we," not some combination of "us" and "them." The struggle against climate breakdown is an expression of life advocating for itself against forces that, because of their own destructive narrative logic, don't find the enormous damage they're doing to be sufficient reason to change course.

Now to the structure of the book. Though my primary purpose in writing is to focus on story and narrative as means of understanding the world, I first use chapters 1 and 2 to plumb the shortcomings of other foundational tools of understanding that humans share, namely, science and numbers, our senses, and language itself as a mosaic of culturally determined meanings. I strive to show that all these tools, valuable as they are, have fallen short in equipping us to apprehend what lies before us. They constitute necessary, indeed vital components of an effective understanding of climate breakdown, but they all have flaws that often as not impede a full view.

As creative and verbal animals we often use story to patch those flaws, to create a comprehensible picture from the raw materials of sensing, measurement, and the linguistic components of language. It is stories that distinguish us from other animals, that most make us human—and it is this gift of artful language that I explore as the heart of my argument in chapter 3, which lays out how various sorts of stories we have created divert us from a full apprehension of climate breakdown. I use chapter 4 to establish that those narratives that we so readily reach for to explain our way out of doing something substantial about climate change fall into a particular pattern that seems inevitable but is instead the sum total of choices that specific people have made, and continue to make, often out of narrow self-interest. This is tragic, quite literally—but it also points toward ways out, as I explore in the final chapter.

This inquiry grows from a number of roots, including many years of work as an environmental journalist, an equally long stint teaching university classes on such topics as risk and science communication, decades of work and play dedicated to becoming a firmly rooted resident of the place where I live, and a lifetime of being fascinated by the tightly bound connections between humans and nature. I frame my own experience of climate change as very much the product of place. By this I mean in part that I state my case very much in the context of the United States, the nation that has done the most to cause the climate crisis and one where climate change denial remains a potent political force[13]; readers in other countries will have to determine how closely my perspective resonates there. But I also mean the much more specific place I have chosen as *home*—on a high plateau in the American Southwest, where the effects of change are evident in the forms of shrinking rivers, expanding dry periods, and more drawn-out fire seasons—because linking the global to the local is among the best ways to apprehend what's changing. Those local touchstones will appear time and again in my writing as a means of grounding my argument.

In the course of writing, I have realized, too, that my compulsion to explore this topic grows also from how I learned to read closely, or to understand story, as a young man, particularly in my undergraduate studies in comparative literature. A primary lesson I took from that course of study is that any attempt to impose a unitary theory on a work of literature—and perhaps the real world outside the library as well—can easily shatter of its own breadth and brittleness. In other words, what I have to offer you as an author is both my deeply thought-out analysis and an invitation to poke holes in it, because the exploration of knowledge, like any tentative efforts to improve how we live on this planet, must be a community enterprise open to both collaboration and contradiction, paradox and doubling back.

My skepticism about what happens when a theory is stretched too far has also prompted me to offer short alternative viewpoints in each chapter. They are expressions of hope that present a counterweight to what I well know are the sometimes bleak conclusions I draw in the main text. You might think of them as antidotes to abstract arguments, or as exceptions that prove the rule, or even as my own efforts to resist my generalizations. I think of them as luminescent signs reminding us that even when the night is darkest, there are ways forward.

By answering the question *What is the matter with a society that would willingly destroy its own future?* with an exploration of story, I am not implying that there are not other important answers to this question. Foremost among these is the leading role that well-funded interests firmly rooted in the fossil fuel industry, and umbilically bound to powerful political parties, have played in obfuscating and lying about climate change. The degree to which this deception has been carried out for reasons of profit, political gain, and ideological purity is sufficiently self-interested, and sufficiently well informed, that only the word *evil* can accurately describe the intent. A full accounting of how this program of lying for the sake of power and profit has been carried out would take a book, or more. Though I examine some of the narratives shaped in tandem with this effort, this is not an in-depth look at what these enemies of the future have done; fortunately, other authors have written those books and reports.[14]

With this essay I am also not suggesting that engaging with story as an abstraction should, for anyone, substitute for practicing the myriad sorts of everyday action that are important steps in responding to climate change, such as driving and flying less, eating less meat, eating locally, reusing and repurposing rather than buying new, insulating old houses or building new ones with smart design, working to develop local resilience, restoring endangered ecosystems, rethinking agriculture to emphasize carbon storage in soils, and many more. Indeed, my primary point is that searching for an overarching narrative can easily become a diversion that distracts us from taking those basic steps, about which recent years have also brought us many useful guides.[15]

What I hope to suggest by the end of the book is that too often, in seeking to come up with stories about climate change that are comprehensible but remain disconnected from meaningful action, we have missed the point. Story embodies a delicate balance between perception and action. It is the flexible tendon that connects how we perceive the world with how we as a result decide to act, and both accurate perception and well-aimed action are now more needed than ever before. If we're missing either of those elements we hamstring ourselves, ending up with narratives that might be compelling but that do little to address our socio-ecological crisis; in the end, they are a distraction from the work we need to do.

We stand on a knife edge of history, still able to choose a path better than that of inertia, still able to realize that *breaking down* need not be an end; it can, instead, constitute a necessary and positive step that precedes reconstruction. As Robert Jay Lifton, a psychologist who has written of the parallels between the threats of nuclear war and climate change, has noted, "From now on, our actions are always and never too late."[16] Breaking down how we talk about climate change, too, is much too late, and just on time.

1

PREDICTION

We knew, we knew well enough to be made uncomfortable by our knowledge, but we didn't *want* to know. Our awareness was weightless. So weightless perhaps that if we behaved as though it didn't exist, it would go away.
—Jan Zwicky[1]

Imagine trying to quantify all the energy that has gone into discussing, framing, and debating climate change since it first entered the public consciousness in the late 1980s.

I don't mean physical energy or how much the planet has actually heated up in that time; I mean mental and emotional energy.

How much human thought, calculation, analysis, concern, anxiety, panic, frustration, debate, greed, compassion, bloviation, stupidity, grief, and neuron-to-neuron connections of all types have been spent in that time on understanding, accepting, rejecting, arguing about, preparing for, and worrying about climate change?

How much intellectual effort has it taken for scientists to understand the problem, to take all those trips to Antarctica and Greenland and the remote oceans and various outbacks all around the world in pursuit of readings of temperatures and rainfalls and the sampling of ices and sediments, tree rings, corals, cave deposits, the upper sides of clouds and the undersides of ice shelves and glaciers, and then to crunch all those data points into enormously complex models that both tell us where we are and, alarmingly, point to where we are going?

How much has it taken to engage in endless political debate and negotiation on the topic, to organize and talk through the delicate and tedious and never-enough international conferences that we refer to by the shorthand of their place names—Rio and Kyoto, Copenhagen and Paris—and to practice spirited if often seemingly futile debate about the issue in every nation and in governing units of all sizes, from neighborhoods to the UN?

How much money and calculated planning have gone into the disingenuous and self-interested sleights of hand of the organized climate change denial movement, into the strenuous efforts of what we might call the corporate-ideological complex to claim that climate change isn't happening, or is happening but isn't a problem, or might be a problem but isn't linked to human activities, or is a benefit rather than a threat, or is a golden opportunity to make bank because in a crisis people will pay almost anything to feel they are being saved, even if by the same interests that led them there in the first place?

How much gimlet-eyed and well-compensated planning have insurance companies and military strategists and investment bankers done as they assess the problem, out of the glare of media lights, so that they can position themselves and their interests appropriately in what they recognize as a financial and logistical problem and an opportunity of enormous proportions?

How much well-intentioned figuring and marketing savvy has it taken to fuel the countless efforts to engage the young and the old and everyone in between on the issue, to teach children about climate change without filling them with despair, to devise and disseminate lists of ten or fifty or a hundred ways to save the planet, to replace incandescent light bulbs with CFLs or LEDs, to conserve water, to eat more plants and fewer animals, to drive and fly less, to accept that everyone who lives a consumerist lifestyle is partly to blame?

How much adrenaline and sweat and blood have been spent on the front lines by paramedics and firefighters and search-and-rescue personnel as they confront the latest out-of-control wildfire, the river overflowing its banks, the neighborhood blown apart by a hurricane—to say nothing of the physical and emotional toll on those living through extreme weather events?

How much money that could have gone toward other purposes has been spent in countless communities as residents seek, often in vain, to fireproof their neighborhoods, flood-proof their roads and sewer lines, drought-proof their water supplies, heatproof their crops, windproof their roofs?

How much worry and strain and divorce and suicide have been visited on the waterless farmers, the fishers who have lost their work to warming oceans, the subsistence growers and hunters who can no longer subsist, as they abandon their professions or their long-held homes and move elsewhere, often unwelcome, so that their children can have a shot at survival?

And how much anger and face-creasing worry and sobbing and compensatory drinking and antidepressant use and 3 a.m. anxiety and agonizing over whether or not to have children have been experienced by those who haven't necessarily been on the front lines but know enough of what is happening to harbor doubts ranging from the nagging to the existential about the future of their families, their towns, their countries, their planet?

What, in other words, has the emotional cost of climate change amounted to? How can we possibly measure the rivers of ink, the Great Plains' worth of pixels, the gazillions of clicks and keyboard taps dedicated to the topic of climate change?

We can't. So we reach for the term *gazillion* instead, a safely imprecise term that simply means a hell of a lot. The impact that climate change has already had on humans, even before anything like its full physical effects have been felt on the planet's lands and waters and ecosystems, is incalculable in terms of hours or emotions or dollars invested. What number can we assign? Only a ridiculous abstraction like *gazillion*, which is not a number at all but only a way to gesture vaguely at the overall scale of what we know is a ginormous problem.

But let's not stop there. I may sound like the most benighted of climate change skeptics in suggesting this, but imagine that over these last thirty-plus years, humans had as a collective done all they could to meet only the most immediate challenges of climate change—had tried to put out wildfires, rescue victims from floods, help farmers find other work—without knowing a thing about the overall trend. Imagine that we collectively knew nothing about global climate change and how it ties the entire planet together. Imagine that we did not know how atmospheric gases trap energy. Imagine no Earth-sensing satellites, no ice cores, no intricate computer models of how the atmosphere pulses and the Earth system functions. Imagine that as a result, we had no occasion to think or argue about climate change, had not spent the gazillion hours and dollars and clicks and suffered through a gazillion emotional ups and downs. Imagine that global climate change had simply vanished from the human consciousness even as its effects multiplied.

The question is, would we be better off than we are now?

What I am getting at here is not the suggestion that we should all be climate change deniers so that we can agree happily with one another and keep burning fossil fuels at will, complacently denying recognition of every sign of global change until the (surprise!) bitter end, when runaway feedback loops that seem to come out of nowhere make wide swaths of the planet unlivable or irreversibly impoverished for human beings. What I'm suggesting instead is that the tangible benefits of our knowledge so far appear way out of proportion to the effort we have invested in developing and digesting that knowledge, and that we know it, and that this imbalance has badly shaken both our confidence in science as a trusted means of inquiry and our belief that in a democratic society, shared knowledge leads to appropriate policy action. Since the late 1980s we have understood climate change well enough to know that action is needed *now*, and in all that time *now* has ever receded into the future, as if we were squinting from a speeding locomotive at the point in the distance where the tracks appear to meet, sure that contrary to all previous experience we will get there imminently this time.[2] And naturally enough, as we have rolled over every new and failed iteration of *now*, the problem has only grown worse—in fact, of all the fossil fuel emissions produced since the onset of the Industrial Revolution, half have been produced in the past three decades.[3] Since we have known better.

What I am suggesting is that climate change, or our collective failure to meet its challenge, is inextricably tied to the convulsions that have shaken many of the world's democracies in recent years, and the entire structure of world order that arose from the ruins of World War II. Collectively, we know our systems have not been able to deal with what is increasingly apparent as an existential threat, and as a result, a great many people have lost faith in at least one of the following: (1) science, and the whole rationalism-based means of knowing and planning for the future that's been at the core of Western societies since the Enlightenment; (2) the ability of democracies to effectively govern themselves in the face of ginormous problems; or (3) hope, period. Even as the problem and the need for collective action have grown much more urgent, these breakdowns in our confidence in the future have perversely made it harder to summon the needed political will.

Climate scientists speak grimly of feedback loops set off as the planet warms: melting Arctic sea ice that reflects sunlight giving way to dark ocean water that absorbs it, thawing permafrost releasing enormous quantities of the greenhouse gas methane, burning boreal or tropical forests that once stored carbon becoming instead sources of yet more emissions, altered ocean circulation patterns that will change regional weather patterns.[4] Our poor track record on climate change so far has caused its own feedback loop: doubt in our ability to work together, which has had the effect of making it harder to work together.

This can also be phrased as follows. Since the Enlightenment, many have shared a strong collective belief that it is worth learning how to predict the future with greater accuracy because it was assumed that this knowledge would lead to better decisions. But in recent years it has seemed that what we are best equipped to predict is our own paralysis. It's as if all the carbon dioxide and methane and nitrous oxide we've pumped into the atmosphere were functioning as a sort of lens that allowed us, for the first time in the history of our species, to look sort of clearly into the future. The view isn't a good one. Staring slack-jawed like Frodo into the Mirror of Galadriel, we catch glimpses, with awful clarity, of both cascading, inevitable losses and our own impotence in stanching them.

The human desire for a satisfying narrative is intimately tied to planning for the future, to telling ourselves that our decisions in the present make sense because they are tied to a coherent future path. I experienced this in 1990 as I stood on the western edge of Arizona looking at Imperial Dam, a low concrete structure that diverts a substantial part of the Colorado River's flow to cities and agricultural fields in Southern California. I was a cub reporter who'd just gone back to graduate school to learn more about what we at that time still called "the environment." I had just read a book on global warming written by Stephen Schneider, one of the first climatologists to develop a national profile beyond his scientific peers.[5] It was only a year after a bipartisan group of US senators had strongly endorsed cutting the nation's fossil fuel emissions.[6] But the same amount of time had also passed since Exxon, having supported a robust program of research into the connections between fossil fuel use and atmospheric change in the 1970s and early 1980s, had changed course and instead begun to dedicate funding to sowing public disinformation about climate change, and since the administration of George H. W. Bush had decided not to sign on to binding emissions cuts in partnership with other nations.[7] It was two years after the NASA scientist James Hansen, testifying before Congress during a hot summer that saw smoke from giant fires in Yellowstone National Park cause hazy skies above my home in Chicago, had said that the fingerprints of global warming were becoming visible.[8] It was fifteen years since a geochemist named Wally Broecker had published a paper in *Science* titled "Climate Change: Are We on the Brink of a Pronounced Global Warming?"[9] It was twenty-five years after President Johnson had received a detailed report from a select committee of scientists warning that continued greenhouse gas emissions would by the end of the century lead to such undesirable consequences as melting ice caps, rising sea levels, and acidifying oceans.[10] It was thirty years after British climate scientist Guy Callendar, the first to link actual readings of rising temperatures with theories of how changing proportions of gases in the atmosphere could cause global warming, wrote that many of his scientific peers rejected the idea simply because "the idea that man's actions could influence so vast a complex is very repugnant to some."[11] It was almost a century after the Swedish chemist Svante Arrhenius had calculated how much an increase in emissions of the gases resulting from combustion of wood, coal, and other fuels had the potential to warm the global climate; perhaps because he lived in Scandinavia, he thought this might be a good thing.[12] And it was more than 130 years since a pioneering American scientist, Eunice Newton Foote, had reported the results of a study she conducted showing that gaseous carbon dioxide helps warm the atmosphere.[13]

Out there on the Colorado River, the other students and I heard nothing of all this. We were learning about what the writer Marc Reisner labeled the American West's "hydraulic civilization": how modern human settlement of this arid region was possible only because of large-scale retention and diversion of water, especially of the Colorado, whose water more than 30 million people in the Southwest, California, and northwestern Mexico relied on.[14] At the dam I talked to a recently retired Bureau of Reclamation employee wearing a belt buckle thanking him for twenty-five years of service to the agency. He characterized the river—one of the most plumbed and tightly managed waterways in the world well before climate change was a thing to worry about—as treacherous, deceiving, as if it were an unreliable partner liable to sneak off unseen into the desert badlands that surrounded us. "If the water doesn't show up here," he said, "it's piss-in-your-pants time." Managing the river wasn't his job anymore, but still he worried that the increasing numbers of people living in communities reliant on the river were going to strain it to the breaking point. Wealthy Santa Barbara, he was certain, was sure to soon begin shipping in tankersful of fresh water from Alaska.

The sky-blue river ran through burnt-looking desert hills. Slender even upstream of the dam, it was almost nonexistent downstream. I didn't realize yet that climate models would soon be projecting that less water would be available for the Colorado and other arid-country rivers in the Southwest. What I did know, viscerally, was that the river was being shared by more and more people; I'd seen the ever-growing suburbs extending into the desert outside Los Angeles, Phoenix, Tucson, Las Vegas. *Well,* I thought, *this is going to be interesting. Maybe I should stick around here to watch.* And so I did. I was enough of a reporter to recognize that the rationing of the gallons or acre-feet would provide a lot to write about.

The line of organized numbers stretching off into the far distance seemed to lay out a clear story: *Here's where we are; here's where we're going.* From the perspective of Imperial Dam, the story of the Colorado River constituted a set of equations with components such as the average number of acre-feet carried by the river; the demand from California and Arizona compared to the amount of water owed yearly to Mexico; the population growth of the Phoenix area; the volume evaporated from Colorado River reservoirs; and so on and on. These were the pressures that the Bureau of Reclamation anticipated and planned for. And they could all be converted into readymade pieces of environmental journalism:

Arizona versus California, the United States versus Mexico, the states in the Colorado's upper basin versus those in the lower, farmers versus environmentalists, housing developers versus Native American tribes. The tracks went off into the hazy distance, but they didn't veer from the straight population growth trajectory of recent decades.

On our tour we didn't hear anything about global warming. If Bureau of Reclamation planners were considering it, we certainly didn't learn about it from those working at the dam. The limits that seemed pressing weren't the new ones associated with the planet's warming. They were the old ones that had always been there, exacerbated by the steady population growth of the Sunbelt and of Mexico. Nothing new. There would surely be compelling drama in how people played off one another in the future. That would be enough. Like most people, I had not yet wrapped my head around the implications of Hansen's congressional testimony or Schneider's book. Even though global warming was, like the usage stats on the Colorado River, fundamentally a matter of numbers, it didn't yet seem to embody the sort of old-fashioned elements of immediate scarcity and conflict that drove so many stories about resources.

A few years later I was living in Arizona, in position, as I thought, to document what would happen when a growing hydraulic civilization inevitably met nature's limits.

Since the end of the Ice Age, plenty of societies have had to adjust their lifeways to changes in weather and climate. Some have failed or had to move or adapt. The Native American Hohokam people who farmed and moved water around in what is now the Phoenix area built large canals that ended up being precursors to and templates for modern waterways, which often follow the same course. But they abandoned much of their large-scale infrastructure in the 1300s, possibly because of changes in precipitation.[15] At a local or regional scale, climate could and did change, and people had no choice but to respond.

Since the birth of civilizations, though, humans have never seen changes to the global climate of the speed and magnitude that we face now. Whether hunting-gathering groups or agricultural tribes, dispersed rural communities or dense urban populations, today's human cultures evolved through millennia generally characterized by great stability in climate. It was the period scientists know as the Holocene, when in the wake of the Ice Age global temperatures and climate patterns remained remarkably stable in most places.[16] It was against this known and predictable backdrop that we humans developed, with our agriculture, our industry, our rituals, our myriad forms of adaptation to the specifics of place. The regularity allowed people to know when to plant, when to harvest, when to skimp and when to feast, when to chase game or move with the seasons, when to raid or explore and when to hunker down at home. It was the relative unchangingness of climate, its largely predictable flow from year to year, that made us the humans we are.

Let's put it in terms of story. Because it was predictable, climate became a backdrop, a stage set against which human-centered dramas could play out. Once understood as such, it came to be taken for granted. Climate faded into the background to such an extent that it was scarcely conceived of as something to be measured, nor understood as something that *could* change. Weather, on the other hand, did, which is why humans have always had plenty to say and fear about the power and terror and sublimity of nature—but always as a force, or a collection of forces, that acted within more or less understood boundaries, a supporting character rather than a marquee name. "For thousands of years, humanity felt itself to live in environs of essential changelessness, although the smaller shapings of life—seasonality, floods, migrations—were always in flux," the writer Jay Griffiths has noted. "The human mind watched and observed the alterations and rhythmic variations yet set it all in the context of a greater constancy: flux within fixity, mutability within larger immutability. Unpredictable weather within predictable climates."[17]

Climate's docility, its ability to camouflage itself so that it was easy to forget that it had any potential to be a significant actor at all, allowed the real action to take place on-stage—and that action came to be profoundly human-centered. A stable climate allowed the development of layered societies with sophisticated physical infrastructures—like the Colorado River's complex plumbing system—and of a vast array of means of being in and understanding the world: myth, culture, religion, science, economics. Climate was civilization's hidden partner. It afforded humans the luxury of species solipsism, of believing that however wild and bloody the action on stage grew it would not affect the fundamental structure of the theater itself. Just as a child raised in coddling and protected circumstances might become narcissistic, we developed an unquestioning belief that of course, our stories had to center on us. They were under our control. It seemed natural that they should be.

The early indications that this would not always be so came in the hard currency of science, in obscure numbers measuring the concentrations of various gases in the atmosphere. They arose out of direct measurements that a researcher from the Scripps Institute of Oceanography, Charles Keeling, began making on top of Hawaii's Mauna Loa in the late 1950s. What he found, when he graphed the result, was jagged ups and downs that, on an annual cycle, neatly track the respiration of Earth. When it's spring in the Northern Hemisphere, which has a lot more land than the Southern Hemisphere, carbon dioxide levels drop because plants are absorbing it to grow. With the arrival of autumn, the atmospheric concentration rises because those plants mainly stop growing. In the absence of an industrial revolution the graph of carbon dioxide concentrations would look like the world's longest crosscut saw blade, a steady up and down as regular as breathing.

But what Keeling saw when he began his measurements on Mauna Loa wasn't a flat saw edge. The annual respiration of spring and fall was regular, but so was the trajectory of the line, which exhibited an uptick at regular intervals.[18] Every year the measured concentration, spring or fall, rose a bit, around one part per million, with the rate of increase itself rising a bit each year. In 1958 it was about 315 parts per million. The increase was steady; it was predictable; it was explainable. It could be projected out into the indefinite future unless humans dramatically changed their behavior.

Several times over the next few decades Keeling's measurements were almost interrupted as some government office or other threatened to pull his funding. But he persisted. The number ticked upward, year after year, decade after decade. His work came to be known as the "Keeling Curve," a steady upward march in lockstep with humanity's rising prosperity, as reliable as the burgeoning postwar stock market and the global influence of the United States and its allies appeared to be—or, to view it negatively, as easy to foresee as the rising blood pressure of an aging man who eats too much salt and doesn't exercise. This was the mark of the Industrial

Revolution, the otherwise invisible track record of our burning of coal and oil, of our progress. Since 1958 the atmospheric concentration of carbon dioxide has ticked upward from around 315 parts per million past 325, 350, 375 parts per million. It has now exceeded 420 parts per million, and the rate of increase shows no sign of slowing.[19]

When compared to the entire atmosphere, that slight annual bump is trivial: it's the equivalent of about two seconds in an entire week. Two seconds is less time, probably, than it takes you to draw and release a breath. Two seconds in a week—here, hold your breath that long: break the cycle of oxygen in, carbon dioxide out. But only for two seconds. A week later, do the same thing. That's a comparative measure of how many more molecules of carbon dioxide are being added to the global atmosphere this year due to human activity, when you measure them against everything else that's already in the air.

And it's not as if the instruments were measuring a few parts per million of something noxious, or something that might cause cancer. We can't smell it, we can't see it, we can't feel it. Not only is it nonpoisonous, it's a product of our own respiration, and essential to the growth of the plants we rely on. Molecule by molecule, carbon dioxide is harmless; in large quantities, it's essential. No wonder the human cognitive system boggles at the thought. No wonder this seeming trifle of a chemical change is a threat not just to the physical workings of the atmosphere but to the entire apparatus of understanding that we developed over the millennia when the atmosphere was so stable that there was no reason to think about it as anything but a backdrop. No wonder so many people respond to these seemingly most unremarkable numbers with a collective *No, it couldn't be*—if not literally in the form of denial, then figuratively in the form of never allowing climate change to supplant other policy questions at the top of the agenda.

Perversely, global warming numbers are also so big that they tend to float free of earthly meaning in the other direction. The parts per million have accumulated. Those two seconds a week have been adding up week after week, year after year, for many decades. When we started to get serious about burning coal a couple of centuries ago, the atmospheric concentration of carbon dioxide stood at about 280 parts per million. Now, at around 420 parts per million, it has increased by some 50 percent. The total increase has jumped way up from an insignificant two seconds a week. Try holding your breath long enough to represent all that extra carbon dioxide: you'll need two minutes. Try it!

How did you do? Unless you're a trained free diver or surfer who regularly spends a lot of time underwater, probably not so well. Small changes add up. The amount of carbon dioxide we add to the global atmosphere each year, even if tiny in terms of parts per million, is a lot: more than 36 billion tons, to be specific, and that doesn't include the also significant contributions to global warming made by other greenhouse gases we produce, such as methane and nitrous oxide.[20] That's a big conceptual problem. What does 36 billion look like? Do you traffic in billions of anything in your day-to-day life? Most of us don't routinely work with *billion* as a quantity of anything physical or tangible or touchable. *Billion* easily slides into the same mental category as *gazillion*, meaning simply an unthinkably large quantity. We're not equipped to process such large numbers, and tend instead to deal with them by viewing them as bloodless abstractions. Comprehending them becomes a matter of math rather than of emotion.

Psychologists term this tendency "scope neglect," by which they mean a failure to properly sort out the differences between problems that operate at vastly differing scales. The chemist Albert Szent-Györgyi reflected on this tendency during the height of the Cold War when he was considering how people process the threat of nuclear war:

> We talk about an atomic blast but with our brain, adapted to the pre-scientific world, made to handle a primitive fire in the cave, we cannot imagine a fire of 15 million degrees. I am afraid of hot water but 15 million degrees mean nothing to me, nor does a blast, 300 miles away, which could blind me forever. Having been adapted to live within a small clan, I am still touched by any individual suffering and would even risk my life for a fellow man in trouble, but I cannot multiply individual suffering by a hundred million, and so I talk with a smile about the "pulverization" of our big cities.[21]

This challenge of big numbers grows even larger when it comes time to assess the effect of those changed atmospheric gas concentrations; after all, they're a problem only because they store energy. How much? This number is a bit trickier to measure than a simple concentration of a particular gas, but scientists have figured out how to calculate it. The answer:

from the beginning of the Industrial Revolution through 2015, humans contributed 36×10^{22} joules of energy to the Earth system beyond what would naturally have been there.[22]

Say what?

That's right: 36×10^{22} joules. A joule is the amount of energy required to light a 1-watt bulb for one second. If you're sitting still and doing little beyond reading about climate change, you are probably radiating out about 60 joules every second. Body heat joules are why a comforter works on a cold winter's night. We can, literally, feel them. But as a unit of measurement they are far more foreign, more disembodied, than the familiar degrees with which we measure temperature.

What, then, are we supposed to do with the knowledge that in the past 250 years we have produced enough greenhouse gas emissions to store about a gazillion of them in Earth's atmosphere and oceans?

These sorts of numbers are not ones that we can place in the same mental category into which we put the numbers that we understand as regulating our lives, such as: I pay $1,100 a month in rent; at $5.95, that spiced latte is a bit spendy; that basketball game is almost tied; my child's temperature is 102°, so he'll be staying home from school. Climate change happens only when the numbers that measure it grow incomprehensibly large, so it's no wonder that we place it into a mental category separate from our day-to-day lives. But when the numbers associated are placed in more comprehensible form, they decline toward the too ordinary. For example, since 1880 the globe has warmed by an average of 1.1° Celsius. In the context of our everyday lives, that's a trivial number, hardly worth noticing while deciding which coat to wear out the door.

This is the central math conundrum of climate change: an unthinkably large accumulation of vanishingly small events results in changes that, because of their scale, can be considered either trivially tiny or immeasurably large. It has provided ample opportunities for plain lies that take advantage of both ways in which the numbers are incomprehensible. It has long been a stock argument that human emissions of gases are too trivial to amount to much in the Earth system—the "too small" argument. At the same time there's been an equally potent argument that it is precisely because the numbers are so large that we cannot change course—the "too big" argument.

This is a far cry from measuring and fighting about the gallons or acre-feet of water in the Colorado River. These are numbers grown so detached from our everyday scale of understanding that they often function more as barriers than as aids to understanding. These are numbers abstract enough that they are easily hijacked for political purposes. These are numbers that have floated so far away from our realms of experience and decision-making that it has been easy to conclude we can safely ignore them for another month, or year, or decade. These are numbers, above all, whose narrative force we have not learned to integrate into the stories of our lives in any productive way.

What we do integrate more readily into our life narratives than such numbers is the evidence of our senses—but our senses too are unreliable guides to understanding the intricacies of climate change. It feels strange to write this as stories about catastrophic fires, floods, heat waves, and aridity pile up in the media in the form of endless tales of loss, survival, the shocking gut punches of nature gone haywire. The evidence, it seems, is incontrovertible. But it has not been enough to get us to fight or flee, as humans have done since time immemorial when faced with an immediate visceral threat.

Our senses are calibrated to the perennial *now*, to the ever-moving sequence of consuming moments in which we find ourselves. They're really good at comprehending the present tense: the dry heat of the wind, the snap of a cool mountain evening, the scent of moisture on the breeze, the danger of an unexpected sound. But this core task is so immersive that we forget what they told us in the past and easily neglect thinking about what they might tell us in the future. The present, we might say, is overrepresented in our perception of our surroundings.

That's never been truer than in our era, saturated with electronic media that are perennially drawing our attention to the new. I don't mean some conservative's predictable tweet, coming each winter during an East Coast cold snap, that the suddenly frigid weather means that scientists' warnings about climate change are an obvious fraud.[23] But analyzing social media posts can tell us a lot about what we think matters. In a study published in 2019, a team of researchers led by Frances Moore of the University of California–Davis analyzed more than two billion tweets sent in the United States from 2014 through 2016, picking out keywords so that they could identify which ones were about the weather.[24] What they found was that most people don't base their assessment of what's noteworthy about the weather on an entire lifetime of experience, and certainly not on anything longer term than that; they tend to base it only on the evidence of the past few years. As temperatures rise, what was once anomalous and worthy of comment very quickly becomes what we so glibly call "the new normal." On about a five-year time scale, the researchers concluded, what was once unusual—the new high temperature, the epic rainfall, rain rather than snow in January—comes to be regarded as typical. We forget what came before, and once that same high temperature has come again a few times we regard it as simply the norm for that time of year. Even the comparisons that are made in the media about temperature and precipitation are subject to the same sort of alteration. In 2021 the National Weather Service issued new "climate normals," or regional calculations of averages of temperature and precipitation. Because they are based on averages of the past thirty years rather than longer periods, they can have the effect of cloaking longer-term changes for those who watch or listen to local weather reports.[25] To use an analogy from the world of analog technology, our memories function like tape recorders that operate continuously but that over time record over what came before, consigning it to oblivion.

Ecologists have observed the same phenomenon—dubbed the "shifting baseline syndrome"—happening when people assess what is "normal" or "healthy" about plant or animal populations, or entire ecosystems.[26] It doesn't take long for people to forget what they once held as normal. *The new* very quickly becomes simply *the normal*. How it fits into the long term—which is where the term *climate* properly comes into play—we easily misperceive. When the measuring tool we use is our own senses, we are apt to misestimate change that is taking place.

The team's analysis also found that what people find notable enough to tweet about is not necessarily what's most important from an ecological perspective. As you'd expect, they found that people often commented on noteworthy blips, diversions from the expected: heat waves, cold snaps, severe storms. What they didn't tweet about were subtler changes that don't register with our senses as extremes—for example, higher low temperatures at night or in winter, which, according to climatologists, are among the most marked signals of a changing climate, and which according to ecologists can be enormously consequential. In many coniferous forests, it is frigid temperatures on winter nights that constitute the greatest check on the survival of bark beetles. When sufficiently cold nights don't arrive, the beetles multiply rapidly, stressing and killing trees and leading to die-offs and increasingly severe forest fires.[27] Yet the higher low temperatures that are to blame go largely unremarked, in part because they occur in the dead of night and in part because they're the sort of change in climate that most people like.

We forget. We forget what the past was like. We overemphasize certain sensory inputs while neglecting others that we either don't find important or simply like. We imagine that there's a purity to our senses, that they paint for us an accurate, values-free image of the world, but in fact the stories that we tell ourselves color what we see and feel. Faced with the same conditions—say, a summer heat wave—a liberal and a conservative are quite likely to shape from them different perceptions. They will feel literally the same raw stimuli—sweat, the sun's pricking on the skin, a bodily desire for shade—but their brains will process this information in different ways. The liberal concludes, *This is unusual heat—something new*. The conservative concludes, *It's summer; it's supposed to be steaming*. Later, rather than presenting them with the numbers the Weather Service recorded; just ask them whether they found the summer unusually hot. You may well get differing responses.[28]

We are also simply out of shape when it comes to exercising our senses to read the natural world. I am excluding many farmers, ranchers, professional naturalists, and others who in some immediate way gain their livelihood from close observation of what's going on outdoors, as well as those scattered aficionados who find interest and satisfaction in tracking birds, insects, game animals, the timing of flowers, and the myriad other phenomena that fall into the now esoteric niche of "natural history." Besides those exceptions, we on the whole live in a place and at a time when exposure to nature—or paying attention to it—has been on the wane. Where people once calculated what to wear and whether to carry an umbrella—or what to plant, where to hunt, when to prepare for a big storm—through some combination of lived experience and a close look at the sky, a sniff of the wind, most of us now do our calculation by clicking on our phones, which dutifully provide us an hour-by-hour breakdown of what to expect. We don't even need to look outdoors anymore. And thanks to GPS, we have the remarkable ability—unique in human history—to get where we want to go without knowing where we are.

Is it any wonder that our ability to accurately use our senses to understand our surroundings has atrophied? We live in an era when editors have been removing words such as *acorn* and *thrush* from young people's dictionaries and replacing them with *broadband* and *MP3*.[29] Our frame of reference has shifted with great speed from a primarily ecological one a few generations ago to a primarily technological one today. As Jay Griffiths observes, the faith that we once placed in a stable climate system we have transferred to faith in a stable technological-economic system: "Threaten that system," she writes, "and many people feel their minds under threat, which accounts for the fury directed at climate change activism."[30] No wonder, then, that we so easily fall prey to misdirection from those who have reason to cast doubt on the evidence that our senses are in theory pretty well capable of providing about the natural world.[31] For untold generations our ability to process fine-grained information about place has been among our greatest tools in figuring out how to live in an amazing variety of habitats. Now, with that set of skills atrophied, all of us—even those who aren't moving at all—risk becoming climate refugees who, in lacking the ability to understand the language of place, are unable to protect it.

This has implications for both sides of what has become a profound political divide about climate change. Even today many conservatives are apt to reach for any excuse that allows them to avoid admitting to its reality or scope. If western US forest fires are getting worse, that must be because more people have built houses in the woods, or because people and ecosystems today are paying the price for poor management decisions in the twentieth century, such as dramatically increasing fuel loads by extinguishing all fires even in forests that have an ecological need to burn. These alternative explanations have the advantage of being at least partly true. President Trump may have been guilty of willfully ignoring climate change, and was excoriated on social media, when he exhorted Californians to do a better job of raking their fire-prone forests.[32] But it is true that Indigenous people living there had long traditions of using prescribed fire as a maintenance tool—traditions that, if they were broadly reinstated today, would help to control fuel loads and lessen the risk of out-of-control blazes. The same pattern holds in the upland Southwest, where fire-prone ponderosa pine forests are today much more subject to high-severity blazes than in the past, owing *both* to past land-management practices and to the heating and drying trends linked to global warming.[33]

Many liberals, on the other hand, increasingly point to climate change as the explanation for almost anything new and undesirable in the environment. Here's an example from where I live. In northern Arizona, rain typically comes in summer in the form of the North American monsoon: baking early summer heat draws subtropical moisture from the Gulf of Mexico and the Sea of Cortez to the inland deserts and mountains, where it falls in dramatic thunderstorms. The monsoon is a welcome relief, a literal breath of fresh air. It's a cooling break from the hot and gusty days of dust and wind-chapped skin, a chance to stop having to water the garden. Out in the woods things happen. The forest grasses green up. Mushrooms emerge. Wildflowers bloom. The ponderosa pines, drinking and growing, smell of caramelized sugar.

My writing of this book spanned two summers, in both of which the monsoon weather pattern hardly showed up around Flagstaff. It tantalized a bit, forming clouds over the highest peaks, here sketching charcoal virga streaks against the lowering sky, there dribbling a disappointing trace of rain that wet little more than the surface dust, even pelting part of town with leaf-shredding hail. But the deep soakings that we've come to expect as the generally reliable result of weeks of unpredictable storms failed to materialize. Our squash plants withered. There would be no wild mushrooms to collect. The springs at the base of the mesa withdrew back into the earth.

In town people talked of the weather, as we always do. It was a regular theme at the Sunday morning farmers market. *Another sign of change*, some said, ominously. *We can expect more of this*, others warned. And, most conclusively: *Climate change is here. It's obvious.*

But in fact, scientists don't understand well how a warmer global atmosphere interacts with the North American monsoon. The signals are mixed, the trends unclear. It is easy to assume that what we dread—the dryness that stokes wildfires, limits water supplies, crisps the vegetable garden, makes life harder—has to be the result of climate change. But the best available climatological knowledge suggests that the monsoon is not going away. Some measurements indicate that it may be showing up a little bit later than it did in the past, but the trend is murky, as the monsoon north of the US-Mexico border is inherently highly variable. More obvious is that it appears to be characterized by an increasingly "flashy" pattern of rainfall, in which rain is apt to fall more intensely when it comes, while the dry pauses between wet spells grow longer.[34] There are plenty of ecological changes happening in the Southwest that *are* clear signals of climate change: higher temperatures overall, lower river flows, shorter winters, longer dry spells and wildfire seasons, greater stress on forests.[35] Climate researchers have excellent evidence for claiming that anthropogenic influences to temperature and precipitation are responsible for half of the "emerging megadrought" that has prevailed in the Southwest since 2000—one of the driest periods in more than a thousand years.[36] But science provides no clear sign that climate change should cause the monsoon to simply not show up in strength during any given summer.

The climate scientist Mike Hulme suggested in his 2009 book, *Why We Disagree about Climate Change,* that the phrase really refers to two very different things. The first, which he suggested writing in the lower case, is the physical phenomenon that has also at times gone by the label *global warming*. The second, which he wrote as Climate Change, is a social construct entangled with ideology and assumptions.[37] We're not so good at apprehending climate change directly because of all these limitations in processing numbers and what our senses tell us. But we're very good at apprehending Climate Change as an explanatory tool and as a label of political affiliation—as a marker of what group we belong to. This sense of belonging is powerful enough that it inflates perceptions on both sides of the ideological spectrum. If conservatives reach for ever more implausible explanations for why climate change is not happening, or is not much of a problem, some progressives are apt to perceive it as what the Scottish environmentalist Alastair McIntosh calls a "kind of projective anxiety": "climate change becomes a lightning rod both for the anxieties that it generates, and for other anxieties that people carry in these dislocated times. . . . When I listen to the way that frightened people talk about global warming it can seem as if the climate can become what psychotherapists call a 'chosen trauma'—a focus of meaning that objectifies and seems to make sense of wider constellations of anxiety in their lives."[38]

We have to sketch such stories of meaning precisely because climate change never exists in a single moment. It is an artifact of a trend that exists in the connections *between* moments. Arising only out of the relationship between countless individual data points, it threatens to remain an abstraction even as it changes everything. In a fundamental way, we cannot feel it with our senses at all because they are designed to help us deal with the current moment, not with an endless sequence of moments. Like H. G. Wells's Invisible Man, who could be "seen" only by means of his clothes and the objects he moved, climate change can be perceived only in what surrounds it, in fire and flood, storm and drought. And so it either remains invisible, if we try hard, or is seen obsessively, as a character inherently much more interesting than all the noninvisible folks in the room.

The German sociologist Ulrich Beck predicted as much in 1986 in his book *Risk Society*, which introduced the idea that modern societies center on chronic exposure to risks, especially those created by our own technologies. "Risk awareness is no longer based on 'second-hand experience,'" he wrote, "but on 'second-hand *non*-experience.'"[39] To be aware of the risk of radiation, for example, one needs to believe both in the presence of something that cannot be directly sensed in any way and in the experts and technologies needed to measure it. It is almost impossible for risks of this sort, including climate change, to become as viscerally felt as risks associated with firsthand experience. Instead, their unseen nature easily leads them to be assessed with either denial or fathomless anxiety. Climate change constitutes a sort of intellectual glue that links together scattered moments, but the glue remains cerebral rather than visceral.

As much as our thinking brains might will us to do so, these hurdles make it difficult for many to assign climate change to the list of Things We Really Deeply Care About. When the time comes, we feel in our guts what it means to make a frantic escape from a fast-moving wildfire, or a drenching hurricane, or to face the consequences of drought, but it's hard to extend this visceral response to immediate sensory input to the concept of climate change as a whole.[40] In a very real way it is the wildfire or the hurricane that is experiencing climate change, while the victim's experience centers on the fire or the hurricane itself. "I run to soothe a nightmare in my son's head but I do almost nothing to prevent a nightmare in the world," writes Jonathan Safran Foer in his climate change book *We Are the Weather*. "If only I could perceive the planetary crisis as a call from my sleeping child. If only I could perceive it as exactly what it is. . . . The truth is I don't care about the planetary crisis—not at the level of belief."[41]

None of this would matter much, except that the future matters. A story told by numbers, or through recalling what our senses have perceived, is an inert history, a curiosity, when it only describes what has already changed. Like comparing the blood pressure or cholesterol level of a sixty-year-old with that of the same person at age twenty, the numbers measuring how much we have already changed the atmosphere and the oceans are an important but ultimately irrelevant measure of decline. We can't go back in time to alter our own diet or behavior, and the people of the planet collectively can't go back to correct the mistakes of the past. A blood pressure measurement that's higher than it once was becomes important only in dictating future behavior, and our measurements of climate change become important—become *political*—only when they are projected into the future. And everything depends on what direction climate change trends take from here.

Climate scientists have become masters of predicting precisely that, in the form of countless miniature computerized worlds, or models, that project how climate patterns are likely to change as greenhouse gas levels and global temperatures continue to rise. A 2019 analysis of numerous climate models developed since the 1970s showed that the models were on the whole remarkably accurate in predicting the correlation between increases in greenhouse gas concentrations and increases in temperature.[42] But these numbers too are fraught with difficulty, and not only because, like measurements of past change, they are simultaneously too small and too large to readily resonate with our day-to-day lived experience. The foundational numbers describing greenhouse gas accumulations and temperature rises tick upward in linear fashion—at least when they are graphed across a short time scale, as they usually are. They don't scream with urgency because they don't identify any particular point at which action either is more urgent or carries more potential for leverage. They possess a cool logic that has never interacted well with the hot lights and words of politics.

Numerical trends are simply hard for humans to grasp.[43] When injected into a political system that thrives on quick turnarounds and dramas of the moment, the steady and tedious upward progression of these numbers practically screams out *not today . . . because tomorrow will be only a tiny bit different.* And because the numbers tick upward so predictably, they also suggest that they can be pulled back without too much trouble. In a society that has long held itself to be playing out a drama on a stable stage set, it is natural to believe that a slow and steady trend can be reversed whenever we collectively set our minds to it. Weight on the rise? My doctor has told me what to do about it, and I will . . . after I get through the stress and rich meals of the holidays. Taxes too high? We can fix that . . . if you elect the right politicians in the next election. Computer running a bit slow? Time to have IT install the new operating system . . . but not until after I get through this week's work crunch. The steady and predictable nature of increases in greenhouse gases and temperatures readily fuels a sort of engineering mindset, a conception of the world as machine that believes we can always jump in—once we have sufficient reason to do so—and reverse course.

This is a catastrophic misunderstanding. Climate scientists have predicted the alarming number of ways in which climate change behaves not in linear fashion but instead accelerates and feeds on itself. Melting ice on the Arctic Ocean exposes dark water that absorbs more sunlight, causing warming to speed up. A swath of forest burns, releasing tons of carbon into the atmosphere and leaving a scar big enough that the land dries out, making it impossible for the same species of trees to reestablish themselves there. Ocean acidification threatens coral reefs and shellfish populations, and the decline of habitat and food sources causes the disappearance of fish that feed on them and ripples out into myriad human communities that rely on the oceans for protein. There are many more purely physical examples than these. In a 2018 analysis, an international team of geoscientists concluded that a 2° Celsius rise in global temperature would likely unleash multiple physical tipping points that would make it impossible for humans to reverse the warming trend and plunge the planet into what they called a "Hothouse Earth" future. "Hothouse Earth," they wrote, "is likely to be uncontrollable and dangerous to many, particularly if we transition into it in only a century or two, and it poses severe risks for health, economies, political stability (especially for the most climate vulnerable), and ultimately, the habitability of the planet for humans."[44]

It is inevitable, too, that such changes will interact not only with other physical responses but with human responses to them. What's the societal impact when millions of Bangladeshis or Nigerians are displaced because of rising sea levels, or when hundreds of thousands of coastal Floridians are unable to sell their homes? When farmers in arid regions can no longer raise the same crops? When temperatures increase to the point that many densely populated equatorial regions will be literally unlivable for several months a year? When agricultural breadbaskets dry up? When the Colorado River no longer holds enough water to satisfy all its tens of millions of users? These effects will ripple outward from the well-understood root causes of greenhouse gases and temperature increases. The coronavirus pandemic, in which the biological and public health impacts of a new disease rippled outward through national policy debates, the global economy, and personal lives, was just a foretaste of the linked environmental-societal convulsions to come. No wonder that a 2019 report from an Australian think tank concluded that at a total warming of just over 2° Celsius, "more than a billion people may need to be relocated and in high-end scenarios, the scale of destruction is beyond our capacity to model, with a high likelihood of human civilization coming to an end."[45]

Prediction, then, becomes a difficult science as such impacts accumulate, even as climate science models become better and better at explaining how the Earth's systems function and interact. Though the heating of the globe can be measured and expressed as a linear change in numbers, the impact and interaction of downstream effects of heating are going to be anything but linear; they are more likely to exhibit exponential growth, with cascading consequences and the sort of out-of-control timeline that characterized the coronavirus outbreak.

That alarm has repeatedly been sounded, but often in terms—in numbers—that themselves represent a hurdle. "We have just ten years to avert a major catastrophe that could send our planet into a tail-spin of epic destruction," read the website for Al Gore's 2006 film *An Inconvenient Truth.* An internet clock developed by the New Economics Foundation during the same year ticked its way down, second by second, to a climate meltdown that was predicted for December 1, 2016.[46] Even the 2018 report of the Intergovernmental Panel on Climate Change picked a highly specific date, 2030, as the target by which time the world's energy economy would need to be drastically reset to avert disaster; according to some of the media coverage about the report's findings, it was "2030 or Bust."[47] Even though they are based on excellent science, these sorts of prognostications end up bearing an eerie resemblance to the predictions of the end of the world regularly made by some religious sect or another, those uncannily precise math exercises that announce that the end will arrive at 6 p.m. (Daylight Saving Time) on May 21 or some other precisely calculated time worked out through careful study of the Bible or ancient Mayan texts. We have blown through a lot of deadlines, and though local- and regional-scale apocalypse has come to many, on a wider scale it has not, rendering the entire process of predicting the future highly suspect to many.[48]

To have scientists rather than religious überbelievers in the position of saying that the sky is, if not falling, then evolving into something dangerous has thus far largely fallen far short as motivation for climate action. Each time the public hears that disaster is imminent and then finds itself alive and kicking the day after the deadline has passed, the credibility of science suffers another decline. Scientists, it appears then, aren't much unlike any other community of belief; their conclusions are not of a fundamentally different nature than those of economists, theologians, entrepreneurs, politicians, or anyone else.

Science once presided over numbers with regal authority. But as measured in public policy outcomes, and to a significant extent even in public understanding, it has not much mattered that there is an overwhelming consensus among scientists about both the causes and likely consequences of climate change. To scientists the broad outlines of the story are fairly simple: climate change is here; we caused it; the trend numbers are clear; its effects will be dire. But to many others this messaging has become just the dialect of another interest group, one narrative among many rather than the unquestioned backbone of a single shared understanding. And so we can compare the exponential growth of climate change studies published in the scientific literature to the more linear

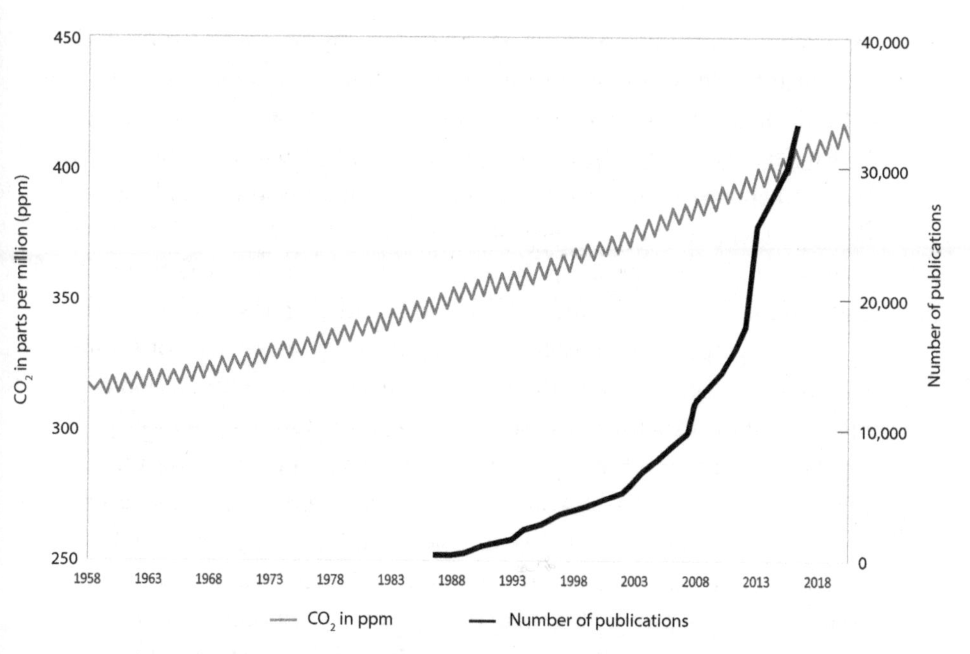

1.1 The Keeling Curve, or atmospheric CO_2 concentrations, measured at Mauna Loa, Hawaii, 1958–2016, graphed against the proliferation of peer-reviewed climate change literature, 1986–2016.

increase of carbon dioxide in the atmosphere as measured on the Keeling Curve (figure 1.1), and find in the gap between those trajectories a measurement of the impotence of science in bringing about needed change.[49]

Hannah Arendt, perhaps the twentieth century's most dedicated student of the morality of public decision-making, pointed out in a 1967 essay that science and other nonnegotiable means of human knowing pose an existential threat to the practice of politics:

> Every claim in the sphere of human affairs to an absolute truth, whose validity needs no support from the side of opinion, strikes at the very root of all politics and all governments. Seen from the viewpoint of politics, truth has a despotic character. It is therefore hated by tyrants, who rightly fear the competition of a coercive force they cannot monopolize, and it enjoys a rather precarious status in the eyes of governments that rest on consent and abhor coercion. . . . Unwelcome facts possess an infuriating stubbornness that nothing can move except plain lies.[50]

With their short-term fixations on policy prescriptions dictated by well-funded interests and looming elections, democracies often have a hard time dealing with long-range planning, especially as the technology-fueled feedback loops between politicians, the media, and the public have grown ever tighter. Under President Trump, this long-standing tendency to focus on the short term at the expense of the more distant future grew to a national pathology—and indeed, the election of Trump and of authoritarian leaders in other democracies can perhaps best be understood as a symptom of mass disengagement with a future that constantly threatens to overdetermine the present. Those leaders have embodied a rejection of the idea that we best engage with that future by getting good information, assembling it to make predictions, and planning accordingly. Make America Great Again sounds like a prescription for the future, but it isn't: in addition to being an emotional jolt that promises fellowship and emotional release in the moment, it functions as nostalgia for a yesteryear whose problems appear manageable (albeit only in retrospect, and only to the cult's adherents) because they took place within the confines of the stable stage set. It's a perspective the writer Howard Mansfield captured in his book *The Same Ax, Twice: Restoration and Renewal in a Throwaway Age*: "The past is a sunlit country morning. The anxiety of today is not present. Just by being the past it implies a happy ending: See they got through, they succeeded, because, after all, we're here."[51]

I understand the impulse. Since I first saw Imperial Dam more than thirty years ago, the amount of water flowing down the Colorado River has declined by about a fifth, and researchers have been able to link that decline very precisely to increased temperatures that increase evaporation and water use by plants. The equations for determining human use of the river have grown more complex. Water managers in Arizona and other states have begun the contentious process of managing for future decline by drawing up lists of who will have to give up how much water at what time. The models posit that the declines will continue—to as much as 30 percent by 2050, depending on how much societies are able to curb their emissions.[52] And it's not just the Colorado. Other western waterways, too, are shrinking, even as groundwater levels drop and dry seasons lengthen.

Staring into these awful projections, I grow nostalgic for what seem like the halcyon days of the 1990s, when disputes about water were—at least in retrospect—simple and predictable. Disagreements between farmers and environmentalists from that time seem almost childishly trifling from today's perspective. I miss the certitude that came of being young and naïve and looking at a future that seemed so much more sure than today's. And I understand why so many people experiencing uncertainty are liable to dismiss predictive tools as those tools increasingly relate what no one wants to hear. Faced with the disconcerting sense that the stage set on which all our dramas play out is beginning to vibrate and shudder in ways none of us have ever experienced before, we confront a future that in many ways looks awful. As Daniel Sherrell neatly phrases it in his book *Warmth*, "the world transforms from a premise into a question."[53] It is no wonder that so many people turn to leaders who offer a short-term release from worry about the matter, promising both that dire predictions about the future must be wrong and that it is OK, indeed desirable, to live in and for the moment.

George H.W. Bush, the first American president to be presented with global warming as an actionable crisis, sketched the template for this political pathway when he said, at the 1992 Earth Summit, "The American way of life is not up for negotiation."[54] If climate change was to threaten mainstream ways of doing things, then it was not to be addressed, period. It had to be wished away, regardless what else might be discarded with it.

Almost thirty years later, Trump completed this circle of inaction. Where other presidents had grudgingly allowed that it might make sense to take some baby steps in response to climate change—though never enough to rock the economic boat—he posited that the entire supposed problem was in fact a hoax, something he chose not to "believe." And just like that, the awful weight of complicity could be lifted.

"Choice is a great burden," Masha Gessen has written. "Totalitarian regimes aim to stamp out the possibility of choice, but what aspiring autocrats do is promise to relieve one of the need to choose."[55] The only choice you needed to make was to believe.

LISTENING TO EACH OTHER, AND TO THE LAND

"Listening in wild places, we are audience to conversations in a language not our own," writes Robin Wall Kimmerer in her magisterial book,

Braiding Sweetgrass: Indigenous Wisdom, Scientific Knowledge, and the Teachings of Plants.[56] This would hardly be news to our ancestors who made their living through hunting, through gathering of wild food plants, or through small-scale agriculture closely integrated into local ecological and climatological systems—through close and mandatory familiarity with the natural surroundings upon which their lives depended. Nor is it news to remnant cultures, like Kimmerer's own Potawatomi Nation, that live closer to the land than most residents of highly technologized societies. But it is news in an era when most of those residents, buffered by complex infrastructures that provide food, water, and shelter, can afford to live in casual neglect of changing natural patterns. Just at a time when a sensitivity to environmental change has never been more important, a larger proportion of the human population than ever before is poorly equipped with the skills to do so.

We might, then, benefit from listening to those who have most mindfully practiced the art of listening. And those who have been best at this are members of peoples who have long experience of particular places and who transmit culture through oral rather than written means. That's because multigenerational oral traditions, when sustained through regular retelling, rely on keeping understanding of the past alive in the present—the past, that is, as an intertwining of humans and place. As the oral history scholars Kathryn Newfont and Debbie Lee have phrased it, "Oral cultures, because they greatly value and actively teach attentive listening, may prepare people especially well to receive and retain messages not only from fellow humans but also from nature beyond humans."[57]

Such accounts of place can reach back almost unbelievably far in time. In his book *The Edge of Memory,* the geoscientist Patrick Nunn documents how aboriginal oral histories from Australia credibly provide accounts of how coastal lands were inundated after the end of the last Ice Age—a process that along that continent largely ended seven millennia ago. The coastal Australian stories, he contends, can be specifically tied to the details of how sea-level rise happened differently in different places depending on local coastal topography. Accounts from the Klamath people about the creation of Oregon's Crater Lake, and from peoples in northwest Europe about ancient cities that vanished beneath rising seas, may share a similar antiquity.[58]

This ability to transcend time and the limitations of a particular generation's lived experience is what gives oral tradition its power. Keepers of such traditions have used them to provide practical instruction in dealing with environmental conditions that no person alive has experienced because oral cultures have been able to embed useful knowledge into stories that show what to do when conditions return to what they were several generations earlier. Oral histories from Australia encode practical information on how to process food plants that appear only every few human generations,[59] while Rory Walshe and Nunn have documented how residents of South Pacific islands have used oral accounts of long-ago disasters to keep themselves safe when a tsunami strikes.[60]

Nunn posits that such stories are deliberately "encoded" into myths starring gods and heroes in order to make them more memorable to future generations:

> Our ancestors were never arbitrarily creative. Everything they did was purposeful because they had to spend so long acquiring the food they and their dependents needed to survive. . . . Storytelling evolved in non-literate cultures to pass on practical information, but in order to convince every new generation of its importance (and ensure that it passed it on to its children), it was made more memorable through exaggeration, identity reinforcement and exciting methods of communication. The pragmatic roots of such storytelling have been lost in most of today's literate cultures, so it is only the other things that remain: narrative, drama and performance for its own sake.[61]

Such oral traditions have unusual if generally underappreciated power in helping to document environmental change by connecting today's conditions to those that prevailed in the past. The geologist Margaret Hiza Redsteer has used oral accounts from Diné (or Navajo) elders to document how the Four Corners region has grown drier since early in the twentieth century.[62] Oral histories my students have gathered from elders in the neighboring Hopi tribe have uncovered a similar trend.[63] In both cases these accounts align with climate change models, but they provide much finer-grained and localized detail. Because Indigenous peoples are often marginalized—in the United States and in many other countries—these traditions represent a storehouse of information that in many places has hardly been tapped.

They also present a pathway to thinking in an intergenerational way that can be applied as readily to the future as to the present. A culture that through retelling and the assigning of value keeps alive the words and observations of ancestors who lived long ago is liable to assign a similar importance to the relationship between present and future, and to resist the sort of prevailing Western devaluation of the future that I explore in chapter 4. That is a tantalizing possibility held out by Kimmerer, who is both a member of a Native American tribe and a scientist with a PhD. "I dream of a world guided by a lens of stories rooted in the revelations of science and framed with an indigenous worldview," she writes, "stories in which matter and spirit are both given voice."[64]

2

METAPHOR

The apocalypse isn't an event; it's an environment.
—Jamais Cascio[1]

For more than thirty years a photographer named Peter Goin has been visiting Lake Powell, the giant Colorado River reservoir that straddles the Arizona-Utah state line. It's changed. In the late 1990s it was a vast expanse of blue set into rust-colored sandstone, full to the brim with two years' worth of the river's flow—nearly enough to provide for the annual residential needs of the entire nation. Twenty years later half that water had vanished into the thirsty sky, leaving a hundred-foot-high "bathtub ring" bleached white on the sheer rock walls. Peter, ever fascinated by the marks people leave on western landscapes, documented the change year by year. Using a large-format camera, he used what he terms "visual poetry" to record what he thought of as the unconventional but evocative beauty that appears when what was submerged is revealed: sunken boats, lost fishing lures, dirt-stained tents and camp furniture. It was, he thought, a new sort of beauty. So when I got the chance to write a set of essays to accompany his photographs I knew I wanted to incorporate a sensibility reflective of the images.[2]

The book was to be an exploration of how to understand a landscape that's changing rapidly in often unpredictable ways. What becomes of a gorgeous if artificial reservoir when it is unlikely to ever again be, as planned, a fully topped-off watery playground? And how might we use the liminal uncertainties of this land- or waterscape as an encapsulation of how all the world's land- and waterscapes are going to evolve in the face of climate change?

What I needed was a good metaphor. And because I was writing about a river, one came readily to hand. Our adventure in navigating a new Southwest characterized by diminished and less predictable water supplies, I wrote, is akin to the adventure faced by the first documented expedition down the river, led by Major John Wesley Powell in 1869. I found the metaphor apt because the canyons of the Colorado really were terra incognita at that time; Powell famously described the descent into the Grand Canyon as a trip into the "Great Unknown," which is an apt phrase for where we stand today.[3] Once you start steering heavy boats down a whitewater river through deep canyons there really is only one way out—all the way down the river, regardless of how churning and dangerous the rapids to come might prove to be. And the metaphor was also useful because of the mingled sense it held of adventure and danger. Though Powell leveraged the trip into fame, a successful career, and eventually getting the giant reservoir named after himself, three of his crewmen died after they climbed out of the Grand Canyon, appalled by the prospect of tearing through yet another dangerous rapid.[4] For its participants, the Powell Expedition's core task was an existential threat—as climate change is now to ever-increasing numbers of people.

All these useful bases for comparison, I believed, helped me establish what I thought was a deep and regionally anchored analogy for how we might approach the new landscapes of climate change. But there's a problem. The analogy exists in the twenty-first century, not the nineteenth. River trips are no longer existential threats to their participants; they're undertaken for fun. To liken a dangerous new era that we haven't deliberately chosen to what today is an enjoyable recreational experience is to lower the stakes. It suggests that coming to terms with our unpredictable future is an optional exercise, something that aficionados might *want* to do—not something that everyone *has* to do, like it or not. My choice of analogy establishes a mental category, and in this case that mental category is one of aesthetics and enjoyment rather than of an existential threat to survival.

That doesn't mean that my comparison is wrong. But it does mean that it is a bit precious. It isn't going to do much to establish the grim weight of climate change as an urgent issue.

Metaphor, as Jay Griffiths has put it, "is a rope by which the mind can swing from one thought to another."[5] Metaphors and other sorts of analogies are among the best tools humans have developed to describe new terrain and new ideas. They are the embracing and evocative yin that cups the hard and rational yang of numbers, a missive aimed as much at heart and soul as at the intellect. Our distant ancestors figured out long ago that analogy can transcend the shortcomings of our senses, confined as they are to our limited bodies. Out there, beyond what we can readily see, are unseen dangers and unrealized possibilities. We may be clever primates, but we're not quite at the top of the food chain. How do we grasp what's out there? By shining a light into the dark wood. By using our powers of language to compare the unknown with the known. *"It's like this,"* our ancestors came to say, countless generations ago, having learned that the concept *like* can be a great comfort when used to describe something previously unknown. It's an anchoring in the familiar. That new food? It tastes like chicken. That new place? It's similar to where we live in this way, but different in another. *Like* is something you can rest on while you absorb the new. It's sturdy, reliable. To understand something new through metaphor is a comfort, like a child's leaning into the familiarity of a parent's embrace.[6]

But *like* can also become a handicap. It can be a dead weight, or a too short rope. In piercing the darkness, a glaring beam of light can hide as much as it reveals. By washing out the nuances, it can trap us in a comfortable myopia. Even as it opens us to new ways of understanding, metaphor can trap us within a restricted range of possibilities. It guides us in *what to see* by training us in *how to look*. This comes to matter a lot when we are facing the challenge of something as unprecedented as global climate change, which presents us as a species with an entirely new set of challenges that have never existed before.

The difficulties begin with what to call the phenomenon we're talking about: climate change, global warming, or something else? During the first administration of President George W. Bush the Republican political operative Frank Luntz was working to elect conservative candidates to Congress—candidates who wanted no constraints put on the fossil fuel industry. His advice? Don't use the term *global warming*. "'Climate change,'" he wrote, "is less frightening than 'global warming.' As one focus group participant noted, climate change 'sounds like you're going from Pittsburgh to Fort Lauderdale.' While global warming has catastrophic connotations attached to it, climate change suggests a more controllable and less emotional challenge."[7] This is overtly political framing, but it's more than mere spin. The terms really do carry emotional weight. When Americans were polled in 2013 about the associations they have with the two terms, they revealed that *global warming* is in fact perceived as a scarier term than *climate change*—it's more likely to provoke feelings of fear about their own futures, or those of their families.[8]

It's easy to dismiss Luntz and his clients as being highly self-interested in their use of terminology—and little interested in solving any actual problems associated with climate change.[9] But our specific use of language matters at a level much deeper than questions of winning elections, or even of setting policy. As George Lakoff and Mark Johnson pointed out in their influential 1980 book, *Metaphors We Live By*, metaphors can operate on at least two levels, one relatively "literal," one more "imaginative" or symbolic.[10] The metaphor that climate change is like an old-time canyon river expedition is on the one hand "literal," in that both experiences encompass great uncertainty and high inertia: once you get going into one of them, you can't easily change your mind and back out, regardless how much or little you know about what's coming. But metaphors also carry a deeper, more symbolic value linked to the *associations* they produce. In the case of a river trip, that is, for modern readers, likely to be a sense of fun, of controlled adventure without the danger that true uncertainty brings. Today river trips on the Colorado are a commercial enterprise, tightly regulated by the National Park Service and ably shepherded by experienced guides. They're recreation, and generally not an existential challenge.[11] It's this time-bound "imaginative" association, more subconscious than the association about uncertainty and commitment, that now colors the analogy.

Climate change is hard to grasp not only because it is not truly discoverable by our senses and because the numbers that define it are hard to fathom. It is challenging to apprehend because of the linguistic challenges embedded in the terms we use for it. Earth's atmosphere, in what scientists long ago began calling the *greenhouse effect,* captures energy from the sun and releases it back into space only slowly. It's atmospheric gases that do the work, or that in this analogy serve as the panes of glass making up the greenhouse roof and walls. Alter their concentration, and we alter how the energy moves. Putting more carbon dioxide and other "greenhouse gases" into the atmosphere is akin to adding an insulating layer to the greenhouse—it serves to store more energy. *Global warming,* then, is what happens when we improve the performance of the greenhouse. As we might put it graphically:

Greenhouse effect → Global warming

Neither of these phenomena necessarily needs to set off alarm bells. We wouldn't be here at all without the greenhouse effect, and global warming is by itself largely neither good nor bad; Earth's climate has oscillated through warm and cool cycles throughout its existence, entirely separately from human intention, desire, or ability to alter it. It's not until we look at warming's corollary effects that most of what we term *climate change* occurs, with all its attendant controversies—because when we worry about climate change we're generally talking about the secondary effects of warming, such as drought, the rising of the seas, changed storm activity, flooding, and a host of other real consequences that in turn lead to tertiary human-focused effects, such as damaged infrastructure, altered agricultural patterns, climate migrations, and more.

We can summarize this explanation of what climate change is in this way:

Greenhouse effect → Global warming → Climate change

Deciding where along this chain of causation to have a debate is more than a political decision. *Global warming* refers to an earlier stage result of a heightened greenhouse effect than does *climate change*—one that is not inherently tied to good or bad consequences for human systems. And this shows how the problems with terminology are not only political but fundamentally conceptual.

Greenhouse effect, for example, compares Earth to an engineered system that is highly controlled and useful. It's true that the physical processes governing how a greenhouse and our planet capture and release energy are similar—that's the literal meaning of the metaphor. But comparing them in this way, the philosopher Xiang Chen has suggested, may predispose us to think that we might deal with these two systems in a similar way—that's the more imaginative meaning.[12] If a greenhouse gets too hot, there are obvious steps to take: open a vent or a door, put a shade cloth over the roof, turn on a fan to pull the hot air out. The problem is manageable. Professionals know how to operate greenhouses. They'll figure it out. As for the rest of us, we don't really need to worry.

It obviously doesn't require too much thought to realize that Earth is not in fact the same as a greenhouse. One of the characteristics of metaphor that we readily understand—indeed, it is the reason we use this linguistic device in the first place—is that we know that the things being compared are in fact *unlike*. Of course a planet is not the same thing as a particular type of building. The scale is very different, as are the origin and the mode of management. But the pernicious thing about metaphor is that even when we rationally are quite aware of the difference between two things or two concepts, the use of metaphor subconsciously frames how we understand the less familiar of them.[13] Metaphors aren't neutral linguistic tools of translation, like methods to convert kilometers to miles; they shape how we perceive reality—not just *what to see* but *how to look*. As Lakoff and Johnson put it, "metaphorical concepts . . . provide us with a partial understanding . . . in doing this, they hide other aspects of these concepts."[14] The greenhouse metaphor nudges our understanding, through its figurative association predisposing us to believe that, even though we know that Earth is not a building, there is still something manageable about its systems. In tandem with the long Western history of placing great faith in the power of engineering and technology to solve problems, this largely unconscious framing is a powerful conceptual nudge that places the atmosphere's increasing concentration of heat into the mental category that might be labeled "problems we can deal with."

The problem with the terminology of *global warming* is a bit different. The term is accurate: when we take the globe as a whole, a heightened greenhouse effect really does warm things up. But as a stimulant to action the term has a few problems. As Earth warms, the effects are not evenly distributed. Some regions, especially heavily populated and media-centric northeastern United States and western Europe, have experienced extreme cold in winter, probably due to alterations in the flow of the jet stream caused by anthropogenic warming.[15] But *warming* is a sort of sensory trigger word. Hear it on a cold day, and the notion that Earth is heating up seems laughable. To process the word *warming* when the evidence of the senses is screaming *cold!* is to engage in a body/mind dichotomy—and it is the body that often wins these arguments, firmly placing the concept *warming* in the category of "stuff I can't quite take seriously right now."

Global is a no less challenging term than *warming*. In the United States, with its strong tradition of exceptionalism, some listeners no doubt discard the notion of a global problem simply because it is global. Sure, the United States has problems—but they cannot be the same problems faced by the rest of the world because our sense of being special, of being one-of-a-kind among nations, is for many Americans a core part of our national identity—and hence of their self-identity. And a much broader problem is that the concept of the global can readily remain an abstraction; it can never *not* be an abstraction. As the environmental educator Mitch Thomashow puts it, "It takes a chain of conceptual leaps and assumptions to perceive that an enormous globe filled with six billion people and several hundred countries has a shared destiny, a coordinated plot."[16]

Of these terms, *climate change* is the one that most effectively does refer to a genuine problem or set of problems because the phrase does encapsulate such real and often dire effects as altered patterns of precipitation, heat waves, storm behavior, and many more. When changed atmospheric circulation regularly forces the winter jet stream down over eastern North America rather than remaining in the Arctic, that really is a change in climate; it's a trend, a repeated pattern, rather than an isolated weather phenomenon. So is the alteration of how storm systems move into semi-arid areas like the American Southwest, or the melting of Arctic sea ice, which causes lasting changes to both the ocean itself and adjacent land areas. These problems really do constitute climate change, a term that ought to be more frightening, and more of a stimulus to action, than *global warming* precisely because it describes alterations in the Earth system that have the potential to immediately affect human lives and livelihoods.

But here too the terminology poses problems. The first is that many nonspecialists really don't have much idea—or choose not to have an idea—what the word *climate* means, or how climate differs from weather. When one of those winter cold spells descends on the East Coast, pundits and politicians seize on it as evidence that warming is obviously not happening. We are so governed by the immediate evidence of our senses that we are really bad at assessing long-term trends. *Climate*, when we try to tie it to our lives of perception, is a murky term to begin with, as prone to remaining abstract as *global*.

The second problem is that even more vague word, *change*. It's accurate enough, but only because it's expansive enough to encompass almost anything. This is a broad opening for politically motivated critique because climate *is* always changing, even if what characterizes our current era is the linked facts that climates are changing much faster and more substantially than humans are used to, and that humans themselves are clearly the cause. Simply, the use of the term *climate change* opens the door to the commonly spoken observation that "climate is always changing." This is a classic rhetorical device, a truth that entirely misses the point. It's a diversion of attention—but rhetorical devices don't become classic unless they work. For a significant portion of the population, the term *climate change* defuses itself by carrying within it an indication of its own insignificance.

None of these critiques is meant to suggest that there are any deliberate inaccuracies in using particular terms. Rather, they are intended to show that our usage of particular words and phrases shapes not only how we talk about but also how we perceive the phenomena attached to them. Collectively, the terms *greenhouse effect*, *global warming*, and *climate change* carry within them enough defusing power that they themselves constitute potent barriers to action. Does this mean we need new terminology? Perhaps it does. And perhaps it means that we need to take a deeper look at the challenges that inhere in trying to wrap our minds around something entirely new.

It is not a novel argument to propose that our language has fallen short in alerting us to the dangers of climate change. Bill McKibben in 2008 described global warming as "essentially a literary problem. . . . A crisis in metaphor, in analogy, in understanding. We haven't come up with words big enough to communicate the magnitude of what we're doing."[17] A few years later, in 2016, he felt alarmed enough to propose that we need to approach climate change by moving beyond analogy entirely. Climate change, he suggested, is a foe that is claiming territory in the form of lost Arctic ice, eroding coastlines, and dying coral reefs; provoking the flight of millions of refugees; and actively killing human beings through fire, disease, and famine. "By most of the ways we measure wars, climate change is the real deal," he wrote. "Carbon and methane are seizing physical territory, sowing havoc and panic, racking up casualties, and even destabilizing governments. . . . It's not that global warming is *like* a world war. It *is* a world war. . . . If we lose, we will be as decimated and helpless as the losers in every conflict—except that this time, there will be no winners, and no end to the planetwide occupation that follows."[18]

Despite McKibben's claim that this suggestion is not an analogy at all, it's hard to read it as anything but, if only because climate change constitutes an amorphous enemy. How would we go about declaring war on molecules? It's not they who are claiming territory but their allies, so to speak—drought, floods, sea-level rise, all of them extensions of natural processes, just as terrorism is no more than an extension of the all-too-human processes of violence, political protest, or psychopathy. It's hard to literally declare war on a global phenomenon that is literally hidden in the air.

To be fair, McKibben's suggestion focuses more on the magnitude of the needed response than on the problem itself, and in that way he definitely is drawing an analogy, and a needed one. In the 1940s, he points out, the United States was able to within months make its ultimate economic recovery from the Great Depression; was able to reorient and resize its industrial economy to focus on manufacturing war matériel rather than consumer goods, quickly upend social mores by recruiting African American and Native American men into the armed forces and women into industrial and other traditionally male jobs, and above all achieve a rare degree of common national purpose by agreeing on the sacrifices needed to defeat shared enemies. This degree of mobilization, he writes, is what's needed to effectively combat climate change.

It's a compelling argument, made more so by the track record of fossil fuel executives and their political and economic allies, who in their willingness to obscure the science and sway policy decisions for short-term economic or political benefit have increasingly cast themselves as a genuine enemy. Surely only a massive reordering, one ensuring that public and private decisions work together to defeat the continued warming of the planet, is commensurate with the crisis we face.

Maybe McKibben's analogy works if we are careful always to specify, as he does, that the struggle against climate change is like a *world* war, specifically. But there aren't many people around anymore who remember what a world war is like, so the metaphor may be losing its once visceral resonance. It is true that in one study, researchers found that framing the struggle against climate change as a "war" motivated audiences to feel greater urgency about it than framing it as a "race" or an "issue."[19] But still the analogy that climate change is a war—one that we are currently losing—has problems. It might be spot-on in its roughly literal meaning, in pointing to how we as a society ought to respond to this all-hands-on-deck crisis. But unfortunately, that meaning threatens to be swamped by the more imaginative resonance that the term *war* has come to carry in the United States. How many wars is the United States currently engaged in? Since the end of World War II the United States has more often been involved in overseas military engagements than not. We are still ostensibly fighting a war on terror and a war on drugs, though here too the order of battle is murky, the strategy hard to grasp, the prospects for any outcome other than drawn-out stalemate dubious. The closest thing to war that most Americans experience are the alarmingly frequent outbreaks of gun carnage—and, though this violence claims more lives in the United States than have the actual international "wars" the country has been engaged in during the twenty-first century, it is one large-scale cause of casualties that is generally not distinguished by the word "war" in mainstream discourse. Whatever else our recent national wars might be, they hardly rise to the level of existential struggles for our nation.[20]

Lakoff and Johnson described decades ago what's happening here. Metaphors, they wrote, do more than offer clues to the structure of some new concept or experience by comparing it to a known one; they also reverberate through what Lakoff and Johnson call a "network of entailments."[21] By awakening and connecting past memories, they wrote, some metaphors serve "as a possible guide for future ones." It's this quality that enables such metaphors as Ronald Reagan's 1984 "It's morning again in

America" slogan to resonate so deeply. Not only did Reagan use his metaphor as a ready shorthand for the existence of new possibilities, he also thereby linked his presidency to a set of deep, even subconscious feelings aroused for many people by the idea of "morning." As former presidential speechwriter John Pollack has written, heavy use of a television ad featuring that line helped inoculate Reagan against criticisms that he was too old to be president.[22]

In the case of the war metaphor, people in the United States might be said to be suffering war fatigue.[23] War has become background noise. War, we might conclude, is simply what an empire does as it seeks to pursue its self-interest in countries around the world. War presents economic opportunity for corporations and personal opportunity for young people who are largely closed out of other, less dangerous pathways. War is political incitement, an expensive means of gaining political support by demonizing an identifiable enemy, whether an actual group of people like the Taliban or ISIS or an abstract idea like "terror." War is less a unifying force or a call to action than another symptom of narrow national self-interest, or of the debasement of politics for short-term gain. Declaring a war on climate change risks adding climate change to the long list of Things We Don't Like; the analogy tacitly places climate change into the category of seemingly intractable issues that we feel very, very strongly about but that ultimately some designated specialists deal with, rather than something that everyone needs to contend with on a daily basis. In other words, we place it pretty much where it is today.

The challenge of figuring out what words to use when we talk about climate change is inextricably tied to decisions about what sort of framing we use to categorize it. The herculean efforts of the fossil fuel industry and related interests to deemphasize climate change are a premier example of what the sociologist Philip Smith once termed a "genre war," meaning a behind-the-scenes tussle that's fought to define how the public comes to understand a policy challenge. Smith, in his 2005 book *Why War?*, used this conception to explain why some potential wars can be sold to the public as worth the cost in blood and treasure while others cannot. The same dynamic is at work for climate change. Fossil fuel interests and their allies have been highly successful at framing climate change as a fallacy or an exaggeration, or as a technical sort of problem that can readily be solved using technology and entrepreneurship—that is, as anything but an existential threat.[24] As I write, in 2021, the core word of the political moment for this is "innovation," used to imply that neither consumers nor corporations nor government need change much to deal with climate concerns.

By suggesting that society's adaptation can be more or less painless, this language firmly places climate change in the category of issues to be solved. Yet the core linguistic problem we face is that climate change is entirely unprecedented and so is not an "issue" at all, as the opioid crisis or our relations with the Islamic world might be construed to be issues; indeed, it can scarcely be categorized or delimited in any way according to the tools we use to debate policy. Climate change is, in the neat coinage of the philosopher Timothy Morton, a "hyperobject," a monstrous and inescapable Thing that is at once everywhere and nowhere, a shadowy presence that in its ubiquity touches every part of our lives, making any efforts to avoid it appear ever more pathetic. "We are inside them, like Jonah in the Whale," writes Morton. "This means that every decision we make is in some sense related to hyperobjects. . . . When I turn the key in the ignition of my car, I am relating to global warming."[25] That is not only because the car then begins to produce climate-altering emissions; it is also because the whole chain of technological progress that led to Morton's car being manufactured, and to his needing it to get somewhere, is intimately tied to a long line of decisions about how humans can and should relate to the world around them. As Amitav Ghosh puts it, "Global warming is ultimately the product of the totality of human actions over time. Every human being who has ever lived has played a part in making us the dominant species on this planet, and in this sense every human being, past and present, has contributed to the present cycle of climate change. . . . The climate events of this era, then, are distillations of all of human history; they express the entirety of our being over time."[26]

Paradoxically, this conceptualization of climate change means that those who have for a long time been denying that climate change is an issue are right, at least in a literal sense. Even as climate change is the most fundamental and pressing problem we have to deal with, a Very Big Thing, a hyperobject, it is not an "issue" at all. *Issue* means that some component of public life can be carved off and treated in a certain degree of isolation from others. Climate change cannot be. To refer to it as an issue, or to provide linguistic cues that place it in a category parallel to other issues, is an impediment to progress, a virtual guarantee that nothing sufficient will be done about it.

The breakdown of a generally safe and habitable global climate is not an "issue" for the same reason that money is not an "issue." Health is not an "issue." Politics and religion are not "issues." They are all core elements of human life, touching everything. Take health. It is likely that every one of the randomly assembled passengers in a New York City subway car or attendees at a climate change conference is paying some attention to some sort of flaw or potential in health, even though the nature of those experiences may range widely: the cancer that flared up or is in remission, a loved one's looming death, impotence, an upset stomach, pregnancy or lack thereof, bad breath, a toothache, a hangnail or new pimple, a just completed visit to the plastic surgeon, the brother or sister or friend who's been shot or injured in a car accident—dozens of individual preoccupations that precede or antecede all the mind's other preoccupations about getting to work, about what to have for lunch, about sex, about sports, and yes, about money. We all spend a great deal of energy dwelling on health, and as a society we spend enormous amounts of money trying to correct its shortcomings or at least deal with its limits. But here too it would be meaningless to talk about "the health issue" as if our thoughts on what to do about the subject could fit into a binder of to-dos to be filed next to, say, whether it makes more sense to cut taxes or increase spending on public education, or what to do about Iran.

Like money, like bodily health, climate change touches everyone and everything. It cannot be contained or encapsulated. It is simply the new reality for everyplace and everyone. Unfortunately, unlike money or health, climate change is also new; in general, the phenomenon that we think of as human civilization (also a hyperobject) has developed and prospered during a time of climatological peace and stability. The inertia of our language reflects that, as we continue to define *glacial* as a word that means not only "icy" but also "achingly slow," even as actual glaciers have been dramatically speeding the pace of their retreat at anything but a glacial pace, as Rob Nixon has pointed out.[27] Shall I compare thee to a summer's day? Yes, please, if you're a poet on a predominantly chilly island just coming out of the Little Ice Age. If you're on the same island four centuries later, with record heat waves melting asphalt and buckling rail lines, April or October might be a more romantic touchstone.[28]

In the spirit of trying to break through this linguistic logjam, I'll now attempt to do my small part. Having established that our senses, our use of numbers, and our very use of words all act as a hindrance, I propose to use the term *climate breakdown* where appropriate in the remaining chapters as I turn to examine how our take on the problem is shaped by the stories within which we live. It's not a new term; as far as I can tell, it was first introduced by the British journalist and activist George Monbiot in 2013.[29] But its use remains limited. Many others—including the *Guardian* newspaper, where Monbiot often publishes his commentaries—have chosen to use the term *climate crisis* instead.[30] That too is an accurate term, but I believe that we are sufficiently mired in enough other crises that it will inevitably be watered down through overuse. As climate activists Wolfgang Knorr and Rupert Read have written, the same problem may arise with the widespread use of the term *climate emergency:* even though an increasing number of people live in cities or countries that have declared a climate emergency, the widespread perception that most public actions linked to that label don't look much like responses to an emergency threatens to cheapen the term and cast doubt on the magnitude of what is happening.[31]

What I appreciate about the term *climate breakdown* is that it accurately conveys the real threats: human-induced threats to the global Earth system are accumulating fast enough that the climate as a support system for certain forms of human societies really is breaking down in many places, and this change threatens to unleash numerous corollary breakdowns in politics, economic systems, and societal relations. The term also hints at how the breakdown taking place is not confined to the realm of physical science,but is firmly rooted in our perceptions, our means of communication, and our psychological states. As the historian Christoph Mauch has pointed out, how humans are ransacking Earth's physical resources finds a very real echo in the depleted, broken-down sense of burnout that seems so characteristic of modern life in the privileged West.[32]

Here, then, is my new diagram of causality:

Greenhouse effect → Global warming → Climate change
→ Climate breakdown

This is dire. But I would not be writing this book if I did not believe that what term we will use to describe what comes of climate breakdown—or what will come next in the causal chain—is still in flux, still in part the product of choices we can make. *Breaking down*, after all, can also be a deliberate act of pursuing understanding, a sort of dissection. It is often only by breaking down old means of comprehending our place in the world, however painful that process may be, that we are able to develop new ones that point a way forward.

"What is certain is that we can no longer tell ourselves the same old stories," writes the French political scientist Bruno Latour in his 2018 book *Down to Earth*, which focuses on potential futures in a world dealing with climate breakdown. "All the wisdom accumulated over ten thousand years, even if we were to succeed in rediscovering it, has never served more than a few hundred, a few thousand, a few million human beings on a relatively stable stage. We understand nothing about the vacuity of contemporary politics if we do not appreciate the stunning extent to which the situation is unprecedented."[33] We might learn a great deal by examining that vacuity more closely, and I return to this topic in coming chapters. But first let's investigate the linguistics of the unprecedented. In recent years it became a trope of late-night comedy and social-media eyerolling to poke fun at some new reactionary effort to ban the use of the phrase "climate change" in some public forum. North Carolina banned state scientists from factoring climate change into assessments of potential future sea-level rise along the state's coast.[34] Florida governor Rick Scott was said to have prohibited state agencies from using both "climate change" and "global warming."[35] The Trump administration pressured or induced federal agencies to remove the words "climate change" from analyses, websites, and press releases, while states soliciting federal aid to help them prepare for future climate change–related disasters had to go through linguistic contortions, pointing to such eventualities as "changing coastal conditions" rather than "climate change" or "sea-level rise."[36]

These incidents have generally been viewed as fairly pathetic efforts at full-bore denial—*if we don't mention the name, it can't be happening*. But we can instead interpret them as hints at a deeper meaning, or at a framing device that provides real insight. Perhaps we can find in this language, or this absence, indications of meanings that might be deeper than what can literally be put into words. For millennia, after all, people in diverse cultures have believed that some names hold within themselves so much power that even to speak them is to risk calling down unwanted attention. Human cultures both real and fictional have seen themselves as trembling subjects of powers so awesome, so feared or revered, that people have known better than to utter their full names—from Jahweh to Voldemort—without a very good reason.

In that spirit of reverence, let's typographically hint at rather than explicitly name what we're talking about: C______ C______. And let's explore what happens if we posit that in this suppression of its full name it is not being denied but rather tacitly acknowledged. Political discourse is rich with diversionary locutions, phrases such as *police action* or *nuclear incident* or *economic slowdown* that are carefully crafted to hint at but also to mask some more accurate but dangerous term. In this vein, conservative politicians often talk about the climate crisis obliquely, by juxtaposing such terms as *freedom, prosperity*, and *American way of life* with *sacrifice, alarmism*, or *creeping socialism*. At this late stage, such usages can come to seem less a sign of outright denial than a signal of fatalism, an acknowledgment that C______ C______ has become an awesome and incorporeal power that cannot be named precisely because it would be unimaginable to oppose it.

Politicians from both major US political parties have made it clear time and again that the American way of life will not be given up on their watch, even if the cost is no less than the future of the planet. That's an epochal lack of imagination. It also might be read as an admission that C______ C______ is here and is so powerful that it can only be ranked with the supernatural. Any opposition to it comes at the unacceptable cost of having to give up God-given worldviews. For that reason, it is only going to keep gathering strength until all options for opposing it grow less substantial—and hence less threatening to the worldviews—than whispered prayers.

If we turn denial or avoidance on its head in this way, we might at last find a more appropriate linguistic tag for C______ C______. Our own Frankenstein, it is fast growing in power and, as it sets off feedback loop after feedback loop, threatening to wrestle entirely free of our grasp; before long, we fear, it will irreversibly melt ice sheets, thaw the permafrost, swell and sour the seas, interrupt ocean currents, and raise the temperature of equatorial regions so that they become unlivable. What climate breakdown will do, in other words, is to rear up into the ultimate hyperobject, an unseeable force that is everywhere and nowhere all at the same time; that exercises awesome and unpredictable powers that appear to us to have no limit; that exhibits an insatiable bloodlust for sacrifices whose ethical weight we cannot assess according to mere human frameworks of meaning; that shapes future generations in ways we can but dimly imagine; and that in every other way makes a furious mockery of any pretense of human control of our destinies even though it is ultimately, and ironically, the product of human culture. For many conservative politicians, predisposed by their alliance with the Christian Right and with blood-and-soil nationalists to understand the idea of entities with mythic power, but also nurtured throughout their entire lives on the heady rhetoric of society's control of nature, of endless growth and prosperity, it is entirely reasonable to respond with linguistic subterfuge, with what looks like dismissal but is at heart a surreptitious obeisance to That Which Really Cannot Be Named.[37]

Consider this, then, as the great reactionary contribution to our understanding of climate breakdown, a powerful linguistic insight that comes from the quarter least expected. For it allows us to conclude that human languages have in all their time on this planet created only one locution that can fully encompass what climate breakdown will be. It is the very same one no doubt uttered as a last word by many as they are incinerated in a stalled car as they try to escape a wildfire, shot in some new internecine war caused by unprecedented drought, or swept away by a Category 5 hurricane: *God.*

WHERE THERE'S A *WENDE*, THERE'S A WAY

Egon Krenz was emphatically not looking to change the course of history in October 1989. The new general secretary of the East German Communist Party, and hence the new leader of the country, gave his inaugural address at a tumultuous time. In the Soviet Union the winds of perestroika were blowing, while in Beijing authorities had bloodily suppressed protesters in Tiananmen Square. The longtime East German leader, Erick Honecker, was in declining health and had just been forced into retirement. Each week citizens massed in giant demonstrations in multiple cities, chanting "we are the people" to give voice to their unhappiness with the Communist regime.

"In past months we have not taken societal change in our country seriously enough, and have not drawn the correct conclusions in time," Krenz announced. "With today's meeting of the central committee we are going to institute a change."[38]

The German word Krenz used for "change" was *Wende*, which typically means a substantial change of course—as when a country's policies shift, or a sailor tacks to shift direction. As national leader, what he meant was "change within the existing system," within the Communist dictatorship that had prevailed throughout the Cold War. But that's not what he got. Within a couple of weeks even larger masses were demonstrating in Leipzig and Berlin, the border between East and West Berlin was declared open (largely through bureaucratic miscommunication), and the Berlin Wall began to be broken down by citizens wielding pickaxes and sledgehammers. Within months East Germany had a democratically elected government; within a year it was reunified (albeit bumpily) with West Germany to form, simply, Germany.

In popular parlance, Krenz's word quickly shifted meaning to encompass this entire sequence of events.[39] What once was simply common noun *Wende*, meaning any old change, became proper noun "*Wende*," meaning one specific and highly consequential change. The proper noun, whose closest English analog is probably "revolution," summarizes in only a handful of letters the astonishingly quick and nonviolent shift that took place in Germany as the Iron Curtain was swept aside.

It is a happy coincidence that advocates of reducing or eliminating Germany's reliance on nuclear power had already begun, in 1980, using the term "*Energiewende*," or "energy transition," as shorthand for the shift they wanted to see to using renewable sources of energy.[40] They were heavily influenced by 1970s advocacy in the United States for conservation and renewable energy, or what the energy expert Amory Lovins, with far less linguistic resonance than his German counterparts, dubbed "soft energy paths." But the idea remained a specialist one until 2000, when the country's liberal-led federal government instituted policies incentivizing solar and wind power. Even after a center-right government was elected in 2005, its leader, Chancellor Angela Merkel, continued to push the *Energiewende,* though not as aggressively as activists wanted. (In 2011, after the Fukushima nuclear disaster in Japan, Merkel expanded the policy to encompass weaning the country off both fossil and nuclear fuels.)[41]

Most German politicians, like the majority of the country's citizens, don't dispute climate change, which they refer to as *Klimawandel,* and they evince support for climate protection, or *Klimaschutz.* But those topics have not taken up nearly as much public airtime as the *Energiewende* itself. As a result, Germany stands as an example of a country that has figured out how to talk about climate change in a largely proactive way. To speak of the *Energiewende*—or to argue about it, as Germans often do—is to speak of the single most effective thing a highly industrialized nation can do to combat climate change, namely, dramatic and governmentally mandated reductions in emissions from major economic sectors such as electricity production, industry, heating, and transportation.

It has not been easy, and Germany has not achieved some of its own self-imposed emissions-reduction goals.[42] But the country has progressed further and faster in reducing its emissions than have other large industrialized nations. That relative success cannot be separated from the linguistic success of finding a metaphor for action that works at both the literal and figurative levels. The energy transition really *is* a change of direction in policy and economics. But the word *Energiewende* also calls to mind, or heart, memories of how Germans managed to solve a problem that had previously appeared insoluble—and from the grass roots, without violence. Germany's energy infrastructure, too, is markedly more decentralized than that in the United States, with solar panels on houses and businesses throughout the country and small clusters of municipally or cooperatively owned wind turbines twirling away in fields.

It's certainly true that this change has been fueled by progressive legislation (and legislators), by a public less polarized than that in the United States, and by the relative absence of large, powerful interests militating against action; though past generations of German leaders would never had said so, Germany may well be deeply blessed by its lack of large oil and gas reserves. But as we grapple for the right words we would be well advised to consider the considerable positive weight of a single deceptively simple term. Rather than focusing on the negative, *Energiewende* figuratively brims with optimism. By calling forth warm associations of a difficult task already accomplished, it inherently makes the suggestion that another difficult task might be solvable as well.

3

NARRATIVE

We will never get an ending, and no grievance can justify our desire for it. There will be no credit roll, or curtain call, or finalizing conclusion to this slow-motion emergency. The nature of the Problem is to *just continue*, chaotic and inconclusive, sloughing off all the stories we want to tell ourselves about vengeance or courage or salvation. Maybe this is why we hate it so much, why one side reacts in fear and the other in its mirror, denial. We hate it because it has no allegiance to narrative.
—Daniel Sherrell[1]

"We tell ourselves stories in order to live," Joan Didion wrote some five decades ago.[2] It's the potent first line of *The White Album*, a classic of narrative "new journalism" that traced the bright, fuzzy line between American utopias and dystopias in the late 1960s. And she was right: we create story lines in order to lend meaning to our lives.

But more to the point when it comes to the climate breakdown world before us is that we tell ourselves stories in order to die.

In order to die well, that is, or to die with a sense of meaning drawn from the trajectory that has brought us to this point. In our most potent traditions of storytelling it is the end that molds what has come before. Didion knew this, pointing out later in *The White Album*'s first paragraph that we "look for the sermon in the suicide, for the social or moral lesson in the murder."[3] The ending is the gathered calm after the storm, the essential deep breath without which we are left hanging anxious and off balance. It is the one vantage point from which the entire narrative is finally discernible, from which all hint and foreshadow fall into place. It is the wrapping up of the foreordained, the paired railroad tracks coming to meet at last after crossing their expanse of plain. It is this sense of an ending—a *good* ending, which is not at all incompatible with *sad*, or *tragic*—that has always fueled many of humanity's most ambitious endeavors, or at least the ones we ever and ever keep talking about. The good ending is the last stand at the Alamo, or at Masada; it is the heroic succumbing to cancer, the last squeeze of a lover's hand; it is, in its purest Western expression, Jesus on the cross, making things right through his own corporeal end. The good ending is good precisely because its audience can appreciate how the sacrifice grows ever more inevitable as the story goes on, and can then integrate this well-shaped narrative into their own personal quests for meaning.

The essayist Roy Scranton learned this in Iraq in 2004. He was a Gulf War soldier there, going out on daily patrols and witnessing the randomness of the violence he and his comrades faced: snipers, IEDs, suicide bombings. There was no front line, no territory to conquer, no massed enemy to fire at—only a confusing mass of apparent civilians, most of them innocent but a few lethally violent. There was no pattern to it, no way to make sense of who fell and who survived. The only way to understand this chaos, Scranton came to believe and eventually wrote in his book *Learning to Die in the Anthropocene*, was "to accept the inevitability of my own death."[4] Only by every day concluding in advance that he was already as good as dead could he invest the randomness of death in a chaotic war zone with meaning, and thereby find the courage to undertake his daily work.

This sounds extreme, but it is something that soldiers have known throughout history. So have countless others facing even the most heightened sorts of adversity. "The human ability to make meaning is so versatile, so powerful," Scranton wrote in his more recent book, *We're Doomed, Now What?*, "that it can make almost any existence tolerable, even a life of unending suffering, so long as that life is woven into a bigger story that makes it meaningful."[5]

It is this sense of how the ending grows from the rest of the story that poses the greatest hurdle to our crafting any meaningful response to climate change, because the very ease with which our minds glide into the well-worn ruts of narratives that head for a climax makes it virtually impossible for us to grasp that climate breakdown does not have to be that sort of story. We are addicted to this particular form of narrative and to the triumph or glory toward which it must lead. Humans, the literary scholar Frank Kermode wrote in his 1967 book *The Sense of an Ending,* "make considerable imaginative investments in coherent patterns which, by the provision of an end, make possible a satisfying consonance with the origins and with the middle."[6] He meant Western humans specifically, in whose traditions of fiction writing, he claimed, "plotting presupposes and requires that an end will bestow upon the whole duration and meaning."[7] We are so accustomed to following this arc of narrative that we are predisposed not only to view a vast array of stories in its light, but to perceive the world itself in that particular gleam. It is as if in our endless lust for meaning we have converted ourselves into busy machines, taking in the raw material of existence and smelting and hammering and shaping it into a form that is both meaningful to us and readily communicable to others. We have become so accustomed to this work of smithing that when we are confronted with experience outside our well-worn patterns we are apt to ignore it precisely because it cannot be hammered into a shape we desire.[8]

One of the principal benefits of this form of storytelling is that it provides a ready means of finding meaning in the ultimate end that we all face, and that we spend so much time dreading. But as much as we fear death itself, we more greatly fear the idea of death without meaning. Exactly that potential future is easy to perceive in climate change. It's no wonder so many turn away.

The narrative I am writing about is specifically a Western form of storytelling, one that was well known as long ago as Homer, as he composed *The Odyssey,* and was codified in outline form by Aristotle in *The Poetics.* The question that preoccupied these storytellers was simple: What makes a compelling story? And the solution they found was simple too, so fundamental that it has continued to inform everything from classic novels to Super Bowl ads to the great majority of popular movies, from jokes to the way in which we describe a notable happening at work to our friends.

The formula the ancients developed can be readily graphed, with time on the *x*-axis and tension on the *y*-axis. Tension is the glue that keeps viewers in their seats; it is the degree to which we are at any given moment sucked into the story. Tension can be measured in sweat, in clicks, in the audience's pulse rate, in the degree to which viewers are anxiously picking at their fingernails or, if the formula isn't working, checking their smartphones for some diversion. What Homer—or his earlier counterparts around campfires—figured out was that they could get their listeners hanging on their every word by inciting tension early on in the story, and then releasing it, and then inciting it again. By first raising the tension level—*and then the mastodon turned to face me, and lowered his tusks*—the storyteller gets the listeners to make an emotional commitment to the story; they begin to care about what happens to the particular character who is facing danger, or some sort of challenge. As listeners, we find ourselves needing to know what happens next. We make an unspoken contract with the storyteller: *we'll stick with you until we learn how that turned out.*

But Homer was after bigger game, so to speak, than a single hunt; he was crafting an epic. He needed his listeners to stick with him for many hours. A multihour description of a single incident, no matter how gripping, would not do. Listeners need relief from the nerve-wracking experience of tension. After a crisis, they need to breathe normally again. Maybe they'd like to know a character's backstory, or learn something about the time and place where the story is set. Maybe they could use some disarming humor. But the hook was set, and so was the pattern. There needed to be new sources of tension coming into the story. *So Odysseus survives the bloody Trojan War and turns for home. You'll never believe what happens then!* And so, we might put it somewhat flippantly, chapters were invented, or acts, or episodes, creating a structure that when graphed out resembles a range of mountain summits, with the highest peaking out later toward the right side, toward the end of the story (figure 3.1). *Odysseus gets back home, only to face the biggest challenge of all—will Penelope still be true to him?*

Why do those summits need to get ever higher? Because tension is a sort of drug, one to whose effects we quickly grow accustomed. We need to increase the dose in order to experience the same level of hungry curiosity that got us into the story in the first place. And so in the next episode of tension there had better be more at stake than the first

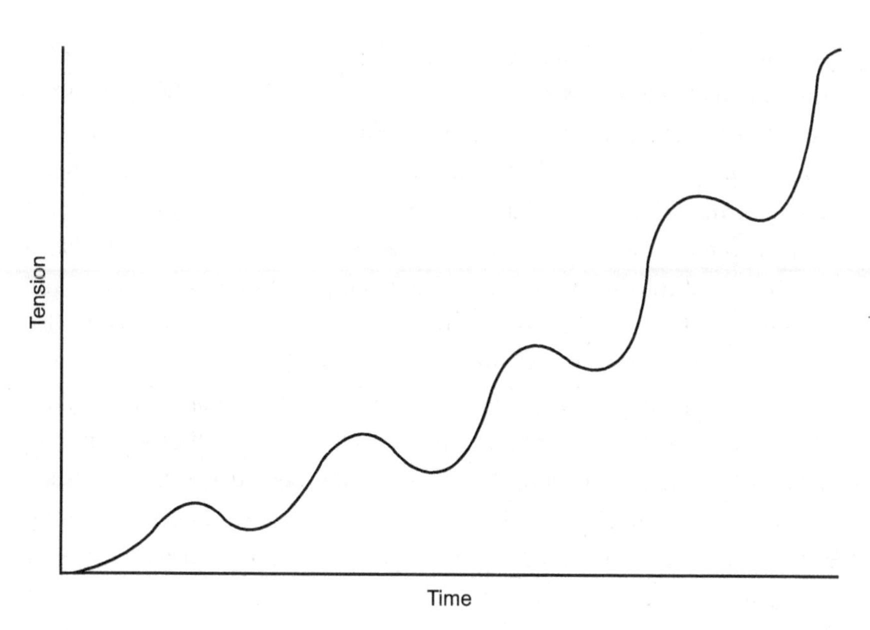

3.1 Generalized graph of tension as a function of time in a classic Western narrative.

time around. If Indiana Jones gets into his first brush with death out of a simple, self-centered lust for riches and fame, there had better be Nazis involved the next time around so that the consequences of failure are bigger—and by the end of the film the wrath of God Himself had better make an appearance.

A compelling narrative, then, requires regular infusions of and releases from tension, and requires that its degrees rise throughout the story, culminating in an ending that begins when the very highest point of tension is resolved. Whether it is happy or not, this is what Western audiences largely conceive of as a satisfactory way in which to close out a story. The various strands of the story line get wrapped up together, more or less neatly, so that as the credits roll there is a retrospective coherence to what we've seen and experienced. What came before is revealed to have led logically and inevitably toward what closes things out.

I want to be clear here that I am writing of a particular Western tradition, but also that this tradition has expanded to dominate popular culture around the world. Other storytelling traditions allow for very different types of narrative, such as epic constructions in which repetition, an overall evenness of tension, and a lack of focus on heroic individual characters—sometimes even a lack of focus on human beings specifically—are perfectly acceptable. Amitav Ghosh explores, in *The Great Derangement*, the tight linkage between the traditional Western narrative form and how individualistic Western ways of thinking and doing have come to dominate the world's geopolitics and cultures—think of all the stories of the heroes of exploration, innovation, commerce, and politics in which doughty individuals overcome adversity to accomplish something that constitutes "progress." "At exactly the time when it has become clear that global warming is in every sense a collective predicament," Ghosh writes, "humanity finds itself in the thrall of a dominant culture in which the idea of the collective has been exiled from politics, economics, and literature alike."[9] These days, what sells in Hollywood is what sells in Bollywood too, and it is this tradition with which we are going to have to come to terms.

It is also important to emphasize that this primary Western mode of narrative has been heavily male-dominated, as Western culture itself has been dominated for centuries by men who have largely wielded control over the great communication technologies of art, literature, journalism, radio, television, film, and social media. This tradition too can be traced back all the way to Homer, at least in how it has been passed down through canons that codify which sorts of storytelling are granted the greatest respect by those who have granted themselves the authority to make such determinations. It is no wonder that stories disseminated through these media tend to be structured in a way that emphasizes male interests and concerns. Ursula Le Guin points this out in her essay "The Carrier-Bag Theory of Fiction," in which she suggests that male traditions of storytelling tend to focus on high-stress individual or small-group acts of exploration and violence, while female traditions emphasize cooperation and working more peaceably within the patterns of nature. It is perhaps no wonder that one of these modalities is often perceived as embodying a greater degree of drama than the other:

> It is hard to tell a really gripping tale of how I wrested a wild-oat seed from its husk, and then another, and then another, and then another, and then another, and then I scratched my gnat bites, and Ool said something funny, and we went to the creek and got a drink and watched newts for a while, and

> then I found another patch of oats. No, it does not compare, it cannot compete with how I thrust my spear deep into the titanic hairy flank while Oob, impaled on one huge sweeping tusk, writhed screaming, and blood spouted everywhere in crimson torrents, and Boob was crushed to jelly when the mammoth fell on him as I shot my unerring arrow straight through eye to brain.[10]

From this analysis arises Le Guin's call for a different sort of fiction writing, one that centers gathering rather than hunting, enclosure rather than penetration. Indeed, exploring different storytelling possibilities that take into account how our predominant traditions are narrowly based in gender and in particular geographies is fertile ground for recentering our understanding of climate breakdown. But I want to step onto more speculative terrain here and posit that our familiarity with stories in which tension steadily rises on its way to a spectacular conclusion taints our ability to address and even to properly perceive climate breakdown far more than is the case with other deep and complex problems that societies face.

Think back to the Keeling Curve from chapter 1 (see figure 1.1). Keeling's graph of carbon dioxide concentrations is a perfect representation of the functioning of Earth's atmospheric and biotic systems as they are altered by humanity's interaction with the global environment—a consistent topography of small peaks and valleys, all in the aggregate steadily rising. If modern industrial civilization were a character, this would be the narrative track of its relationship to what's around it, a steady ratcheting-up of tension (Of course, it is steadier and slower than any Hollywood story could afford to be, which is a big part of why most audiences find it boring). The burning of fossil fuels goes up, the population grows, the standard of living rises.

We're used to this pattern of narrative, this graphic mapping of history, of progress. It rings as familiar, doesn't it?

In fact, we're *very* used to it—even though, despite decades of attempts at public education about climate change and the environment generally, most Americans are not particularly familiar with the Keeling Curve and what it represents. What they are familiar with, in general concept if not in actual graphical terms, is the trend of economic activity over time, the core reason why those greenhouse gas emissions have gone up. This too is easy to graph, at least in the coarse terms of gross domestic product.[11] Since the United States became a nation the growth of the nation's economy—fueled in very large measure by the burning of coal, oil, and gas—has been as predictable over decades as the steady accumulation of tension in the most formulaic superhero movie. This is the summary of our national story, with many of history's most notable episodes—our shared moments of greatest tension—mapped out as peaks and valleys.

We can readily compare the economic growth that took place during this period with the accompanying increase in atmospheric carbon dioxide. The Keeling Curve as a real-time measurement of carbon dioxide has been expanded back in time through measurements of air bubbles sequestered in glaciers and ice sheets. As a result, scientists know very well how carbon dioxide has increased not only since 1958 but for many centuries, even millennia. And when we compare the growth in per capita GDP that the United States has experienced since it became its own nation we see that the rate of growth lines up strikingly with the increase in atmospheric carbon dioxide concentrations (figure 3.2).[12]

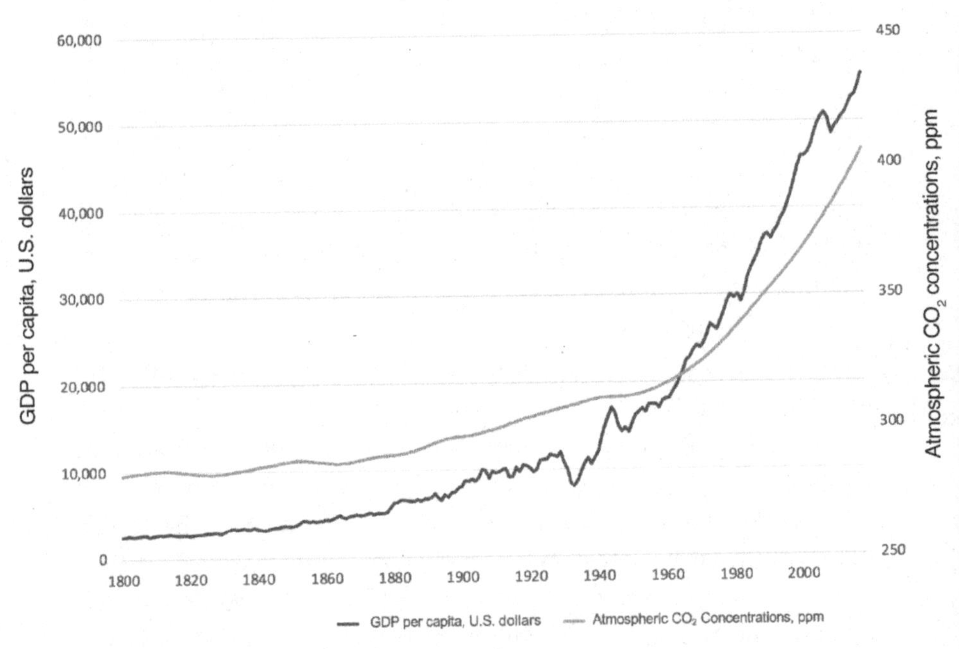

3.2 US GDP per capita, 1800–2018, and atmospheric CO_2 concentrations based on ice core data (1800–1957) and direct measurements at Mauna Loa, Hawaii (1958–2020).

This trajectory of economic growth is so fundamental to our history and to what most Americans believe about our shared future that all politicians know in their gut that to do anything to slow it is to commit professional suicide. No politician ever says *The economy is growing too fast!* Slowing down is an option that cannot ever be on the table not only because it would be equated with economic hardship but because it would run counter to our national self-identity, to our understanding of how our history is to play out—to our belief in the special role that history has specifically designated for us. We are hooked not only on growth but on the *narrative* of growth because continued increase is a core part of our self-identity as Americans. The idea that each generation labors so that the next might have more—more options, more buying power, more freedom—has for many generations been equated with the American Dream. Just as narrative tension is a drug that requires bigger and bigger doses to produce the same satisfactory but short-term buzz, so is economic growth. With every new generation what was once a luxury reserved for the wealthiest—An easy-to-use personal vehicle! Vacation cruises! Air conditioning! Airplane trips to beachfront resorts! Technological monitors of health that peek into every hidden corner of our bodies! A movie theater in every town, in every living room, in every palm!—comes to be taken for granted, an obvious baseline. And so just to continue with our story line, with the meaning we understand our collective lives to hold, we need more. We need more GDP, we need more greenhouse gas emissions, we need to raise the story's stakes yet more. All those graphs line up into a single, viscerally understood narrative of progress (figure 3.3). They could stand as an illustration for the German sociologist Hartmut Rosa's thesis about how modern life is characterized by a widespread feeling that things are ever moving faster and faster: "The experience of modernization is an experience of acceleration."[13]

Most Americans have sufficiently internalized this agreed-upon national narrative, honed and repeated as it has been through generations, that at some level we recognize that to oppose it—by, say, calling for lower emissions at the cost of flattening or reversing economic growth—is not only an enormous challenge but a heretical stance that calls into question many of the guiding assumptions we have shared over

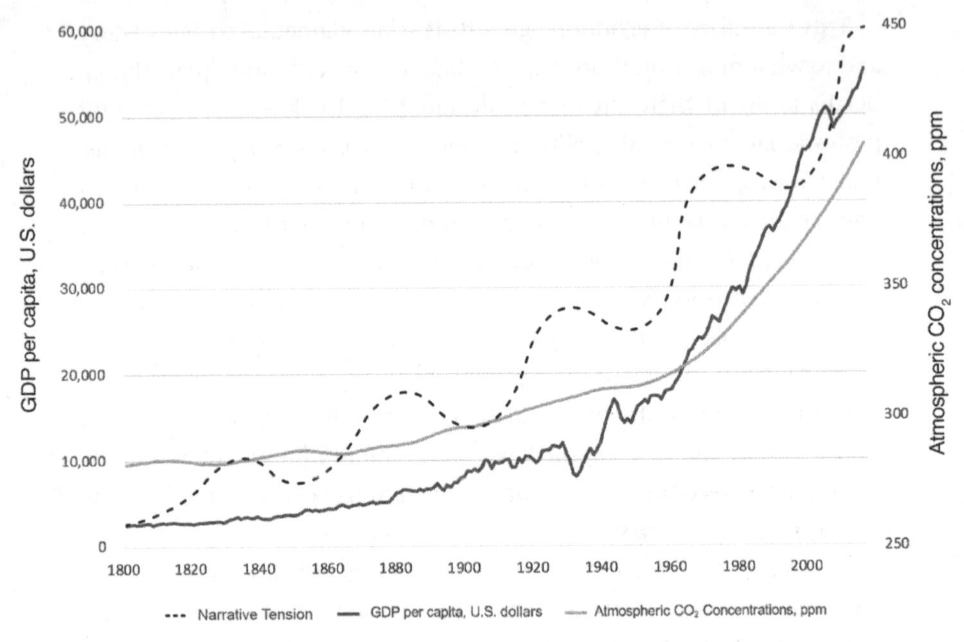

3.3 US GDP per capita, atmospheric CO_2 concentrations, and narrative tension.

those generations. It is a threat not only to the future but also to the past. Reinventing an economy is difficult enough; if doing so means we have to recast our national myth as well, it's no wonder that so many turn away from grappling with the crisis.

At a societal level our ability to respond appropriately, and meaningfully, is also seriously hindered by those who in the increasing signals of climate breakdown find explicit possibility and even hope, for some important and influential interest groups intend to follow the rising curve of tension to an inevitable and satisfactory narrative conclusion that they have decided stands as a core plot point of their own future. Some are adherents of religious traditions who believe that a profound societal test such as that posed by climate breakdown cannot be other than proof of the imminence of God's final plan for humanity—in fact, greater chaos and increasing danger are devoutly to be wished for because they are signs of the Second Coming. The suffering isn't collateral damage; it *is* the point. It is the truth, and the way. In a 2016 poll conducted by the Yale Program on Climate Change Communication, 14 percent of respondents said that global warming is "definitely" or "probably" a sign of the end times. Another 18 percent responded "not sure," while 21 percent chose the only less slightly uncertain "probably not."[14]

The scholar Robin Globus Veldman, who has conducted extensive ethnographic work among evangelicals in the American South, points out that evangelicals' prevailing beliefs about climate change are more nuanced than simply focusing on the end times—in fact, only a minority of evangelicals explicitly embrace climate breakdown in this way.[15] But many others have been thoroughly acculturated by a more than century-long history of conflict between the forces of science and evangelical religion, and more recently by the leaders of the Christian Right, to perceive talk of climate change as a sneaky way to move society toward greater secularization and government regulation, both of which are antithetical to the beliefs of most evangelicals. Or they find the thesis that humans can shape the world as a whole a threat to their core belief in God's decision-making authority. Either way, these believers often ally themselves politically with those who embrace end-times ideas, as well as with other politically and culturally influential climate change deniers. Evangelicals' conviction that working against climate breakdown is a threat to the core of their belief system helps explain why many became such avid supporters of the obviously impious Donald Trump.

But the traditional Christian Second Coming and Armageddon do not represent the only satisfactory end to this story line. As the impacts of climate breakdown multiply in tandem with its saddlemates nationalism, economic inequality, and technological alienation, those with the means and information are increasingly to be found positioning themselves as Chosen Ones with the wherewithal to make it through the tribulations to come. The only beliefs required are, first, that which grows directly from climate science—namely, the certainty that climate breakdown is going to have increasingly dire impacts that will ripple outward from ecosystems into food systems, infrastructure, and social structures—and, second, that the goal of surviving these trials justifies practically any means. The consequences of climate breakdown will be famine, disease outbreaks, attempted migration, war, violence. The rational response? To use one's wits and resources to shore up a piece of high ground. For some it's a ranch in New Zealand; for some it's a literal bunker on the Great

Plains, where Cold War missile silos have been converted into trophy fortresses for the new era of social upheaval; for some it may be literal New Land that is now, if not for much longer, still covered with ice.[16] When President Trump riffed about buying Greenland in his inimicable mashup of real estate developer and Mafia don, it was a sure sign that he had internalized some Oval Office briefing indicating that the strategic frontiers of the New World Order are going to be found in northern lands swiftly undergoing a transition from desolate wasteland to fecundity—if not in crops then in newly accessible minerals and fossil fuels, if not in fisheries then in golf greens glistening in the midnight sun and neatly watered by the runoff from fast-diminishing glaciers.[17] If the lands in question are rock-bound islands easily defended from hordes of desperate, darker-skinned climate refugees from somewhere down south, so much the better; capitalism might then cap its centuries-long narrative of conquest by realizing its final earthly destiny of converting lands once uselessly white with ice to usefully White with the chosen few who have earned the right to be there through the ultimate sign of devotion and virtue—their buying power.

The journalist Christian Parenti has termed this increasingly evident personal climate breakdown adaptation strategy "the politics of the armed lifeboat."[18] It's a scenario that is no less apocalyptic than the Christian one, even if the sorting mechanism that determines who comes out on top is very different. This strategy is clear-eyed in its assessment that climate breakdown is advancing and is going to get worse. It has little to do with climate change denial except that some of its adherents, knowing that the asymmetry of accurate information is a powerful tool for those in the know, spout denial talking points in public even as they privately prepare for what they know is coming. "Although we are unlikely to hear words of this kind in our era," Ghosh writes in *The Great Derangement*, "there can be little doubt that there are many who believe that a Malthusian 'correction' is the only hope for the continuance of 'our way of life.' From this perspective, global inaction on climate change is by no means the result of confusion or denialism or a lack of planning: to the contrary, the maintenance of the status quo *is* the plan."[19] The more the great mass of Americans (and citizens in other nations) dither in denial, apathy, hopelessness, or just plain ignorance or distraction, the better those with knowledge and money and connections can position themselves for the dark times to come.

In this view the most relevant line of rising growth that will inevitably be subject to sudden decline—the denouement after the climax—may not be carbon emissions but rather the global human population. It is a view as consummately self-interested as that of the Nazis—but as perfectly sensible and rational, too, a story with an internal logic and a satisfactory conclusion, as long as you are willing to jettison inconvenient traditions of morality and ethics. It takes the capitalist narrative of endless prosperity for all and crashes it against the rocks—but, as if capitalism were an über-yacht in a James Bond film that eventually has to be destroyed in order to satisfy the demands of narrative tension, at the last moment out pops an even faster speedboat in which the villain, laughing, is able to escape on his way to a sequel. If this outcome bears a more than passing similarity to the many stories that proliferated after World War II relating how leading Nazis—perhaps even Hitler himself!—had escaped to South America, that's an indication that a compelling master narrative that has for some time absolutely dominated a nation's politics and history cannot be abandoned without great cost; however thorough the narrative's defeat, it persists in perverse offshoots that continue to provide some conciliatory meaning to the defeated.

But the lifeboat metaphor itself makes the point that however wealthy this strategy's believers are, however well prepared they believe themselves to be, they are unlikely to find the sanctuary, or the lasting story line, that they want. Lifeboats, after all, are designed for only interim safety and will someday wash up against some inhospitable coastline—or drift aimlessly until the supplies run out. "This sort of 'climate fascism,' a politics based on exclusion, segregation, and repression, is horrific and bound to fail," Parenti writes. "If climate change is allowed to destroy whole economies and nations, no amount of walls, guns, barbed wire, armed aerial drones, or permanently deployed mercenaries will be able to save one half of the planet from the other."[20] That's a rational analysis. As climate breakdown progresses there are likely to be an awful lot of desperate poor people likely to feel that goodwill or obeisance toward the wealthy is a luxury they can no longer afford. But that's a tough realization to come to if you are thickly mired in hubris—if you are a Master of the Universe accustomed to buying all the prosperity and belief you need, say, or if you are the White House adviser Karl Rove, disdainfully consigning journalist Ron Suskind in 2004 to what Rove disparaged as the "reality-based community": "We're an empire now," he said, "and when we act, we create our own reality. And while you're studying that reality—judiciously, as you will—we'll act again, creating other new realities, which you can study too, and that's how things will sort out. We're history's actors."[21]

For people in power who are accustomed to "creating our own reality," the physical feedback loops that rising temperatures are likely to call forth before too long, and the unpredictability of how societies will react to increasing numbers of increasingly severe shocks, are likely to serve as a reminder that when the stage set is burning it is too late for even the most dominant characters to be the authors of their own narratives. The whole premise of the "Hothouse Earth" picture of the planet's future is that physical feedback loops will be powerful enough that humans will not be able to rein in a rapidly heating climate through any conceivable degree of technological innovation, conservation, economic incentives, or international comity. In a Hothouse Earth scenario, the species that for the last ten thousand years has believed itself largely in charge of what happens onstage will find itself with less and less say over the plot.

In the short term, though, it's comforting to believe that the lifeboat is an end in itself. It's enough for now, because a hopeful narrative about the future glows with sufficient meaning in the present that we can approach our highly uncertain times with something like confidence.

Most of us don't assign ourselves to the class of James Bond villains, though, and probably don't see much of a future in some rock-bound Fortress of Solitude. Most of us want to remain where we are—or perhaps move inland just a short way, if we live on a low-lying patch of coastline—and raise our children, and greet our neighbors, and see our communities and not just ourselves out of this mess. Faced with the baroque threat of climate breakdown, what are we to do as ordinary citizens who want life to go on more or less as it has?

Well, according to one highly compelling narrative whose potency serves as another great hurdle in the way of cutting our greenhouse gas emissions, we can do what Americans have always done: innovate and engineer and entrepreneurize our way out of our predicament. By harnessing American know-how and cleverness and hard work, we can get out of our problem with the same tools that got us into it. We'll block sunlight with aerosols dispersed from planes. We'll fertilize the oceans with iron so that they store more carbon. We'll invent machines to suck carbon dioxide from the atmosphere. We'll burn "clean coal." We'll figure it out using the same tools of inspiration and perspiration that have fueled American technological innovation and world superiority for generations.

This is a beautiful story because it represents only the slightest course correction in the narrative trajectory we have experienced so far. It is the most immediately comforting of all our climate breakdown response narratives because it would continue our growth curve of GDP and national progress into the indefinite future. The techno-fantasy of geoengineering is the one narrative that really does Make America Great Again because it relies on the same elements that fueled our industrial and geopolitical ascent in past centuries: ingenuity, ready access to natural resources, and technological prowess. It even has the virtue that if it were to work, it would make its inventors and implementors unbelievably wealthy. It's hard to find a better proponent of this view than Rex Tillerson, who so neatly bridged the country's corporate and political worlds when he gave up his position as CEO of ExxonMobil to become the first secretary of state in the Trump administration. Climate change, he said, is "an engineering problem," one amenable to all the same remedies that clever engineers can use to locate and extract a particularly tricky pocket of buried natural gas: the application of capital, engineering muscle, and a proprietary mix of chemicals, greased by signed or unspoken understandings of mutual interest with whatever political and regulatory powers control the points of leverage.[22]

This is a story line that many would like to get behind, building as it does on Americans' inborn belief in the value of technological innovation and sunny optimism about the future. It would feel so right. We are perfectly primed for it by all the stories we've watched with exactly this story line, tales of the Wild West or the Space Age or a dystopian future in which prospects for success or survival always look dimmest just before the climax, when the superhero (or, better yet, supernerd) or clever band of misfits manages to seize victory from what looked like the jaws of darkest defeat. Just as they do, we will prevail using all the character traits and flaws, all the strengths and weaknesses, that we've gotten to know all too well on our passage along the narrative arc. Our media diet has ingrained deep in us the knowledge that the hero's journey requires a passage through a dark night of the soul, so we can always rationalize ourselves to be experiencing that now as a society or as a nation. And it has also predisposed us to look for happy endings no matter how dark the night grows; even though audiences love heightened tension and cliff-hanger suspense, we prefer it when those narrative tools are marshaled cathartically in the service of an ending in which things turn out OK.[23] We are so accustomed to the happy ending coming out of left field that applying the same hoped-for story line to our real-world situation is a mighty temptation. Seas rising? Storms growing? Climate chaos? Surely we can harness our own inimical qualities of cleverness and innovation and competitive drive—and yes, the greed and back-stabbing and cupidity that go with them—to come up with a surprising but perfect solution, a silver bullet that will cut a clear path through what for now appear to be diverging and chaotic story lines.

Because we are talking here about what we so casually call "the environment," it is also worth noting that a narrative of what we might call techno-optimism has the potential to be particularly satisfying to many of those who, finding environmentalists generally overwrought, are ever eager to remind us all that previous warnings of doom have not been realized. Ever since Rachel Carson and Paul Ehrlich in the 1960s broadcast their warnings about such modern environmental threats as the wanton use of synthetic poisons and overpopulation, proponents of techno-optimism have taken great pleasure in pointing out every instance in which the most dire predictions have not come to pass. Is it any wonder that many of them would read predictions from the IPCC, or Greta Thunberg, with those earlier models in mind? Techno-optimism thus becomes more than a workable response to a genuine physical problem. It scratches the itch for a deeply satisfying political story line because it would, according to its adherents, prove that dire warnings about the future are always wrong—*must* always be wrong—because they do not take into account the infinite capabilities of technology and innovation and good ol' frontier gumption. What greater affirmation than that could there be of the story of American capitalism?

Techno-optimism is a deeply seductive narrative because it appears to equate to having it all. But it only appears to equate; this proposed set of "solutions" to climate change would do nothing to address all the allied problems of economic inequality, racial oppression, or xenophobic nationalism that are so closely tied to why we are facing climate breakdown in the first place.[24] Those most inclined to believe in the techno-optimism narrative, though, are unlikely to want to explore the interconnected nature of all these symptoms too deeply. No need for one of those pinched environmentalist narratives of sacrifice and finding joy in potlucks and the close observation of nature, it says; let us instead continue to find our satisfaction in the same way those who came before us did.

You don't have to be a dyed-in-the-wool devotee of progress and the American way of life to feel this way. Sitting here in a dry part of the world that's getting drier, steeped in the literature of climate breakdown, I often find myself nurturing an obscure hope that someone will find a way out of this mess so that we might avoid the chaos to come. But in the glare of a quickly warming world, that hope does seem desperate because the scale of the problem is so big, the supposed solutions on offer so flawed. Yes, we could block sunlight with aerosol particles so that things wouldn't get so warm, but that would do nothing to stop the acidification of the oceans, would probably result in wars because different countries would find widely differing degrees of benefit or harm in the climatic effects, and, if the practice were ever stopped, would result in even faster rates of warming. Yes, we could capture carbon dioxide from coal-fired power plants and bury it, but to do so is so energy-intensive that we'd have to boost the size of the power plants by a third to fuel the process, and burying it would have its own impacts. Yes, engineers have developed means of capturing carbon dioxide from the atmosphere itself, but will we really have the wherewithal, the money, and, yes, the energy to produce and operate billions upon billions of these machines? As the planet morphs unpredictably into a Hothouse Earth, how will our technology and economy possibly keep up with the accelerating rate of change?

Even the lower-tech means of trying to balance the atmospheric carbon budget look extremely challenging in the face of current political and economic conditions—which is to say, in the face of all these alternative climate change story lines that continually divert us from doing something effective. Yes, it would be wise to conserve tropical forests, and to rethink agriculture so that more carbon is sequestered in soils. But proposals to do this sort of work have been around for decades. They have been laudable for that entire period, and they have in the aggregate been ineffectual. Even as the effects of climate breakdown are amply evident on every continent, forests are still coming down or burning up—and the toxic interconnection of climate chaos, doubts about the efficacy and longevity of democracy, and overt nationalism makes it seem unlikely that expanded global agreements for conservation will be on the upswing anytime soon. Is the climate breakdown danger going to loom large enough anytime soon so that sufficient numbers of people change course?

The jury's still out on that one.

None of this is meant as a counsel of despair. But it is intended to point out that we are not going to find our way out of this predicament according to the familiar story that has prevailed in the modern era, namely, that climate change is a problem we can leave to some experts somewhere, whether engineers or entrepreneurs or religious shepherds. All of us are part of the great personal and collective projects of building meaning through story; a national narrative is the sum of a great many individual and family stories that we learn to understand as parts of a greater whole. Reshaping the climate breakdown story so that we may at last face it honestly is a project large enough that we all need to grapple with it.

And many of us are trying to do exactly that. What does the climate breakdown narrative look like for those who accept that climate change is real, and terrible, and likely to get much worse? What happens when we are unable to subscribe to any of the convenient off-ramp narratives that portray it as not a problem, or as a necessary partner to desired changes, or as a simple issue that can simply be solved?

For many of those who take it seriously, the prospect of climate breakdown has become a sufficiently major part of daily life that it is distilled into the personal every day. This distillation often takes the form of cognitive dissonance, which might be defined as the recognition that even though the world as we know it is ending, even though everything is going be different, we still need to lead our daily lives as before. The seas are rising, and still the laundry piles up. Forests are burning in unprecedented ways, and it's time again to respond to the bulging email inbox. In the Arctic permafrost is puddling into mud but I can't focus on that because I need to enroll my teenager in his first high school classes. Coral reefs are bleaching a deathly white but what I need to think about is how to stock up on groceries: risk a trip to the store while the coronavirus is going around or use the pricy delivery service? Crops are drying up in Central America and on the Indian subcontinent and the Great Plains, but in the meantime my wife and I have an appointment with our financial adviser to discuss whether to put money into our retirement fund or our son's college account. We talk about tax benefits and cash flows and the spreadsheet holds dates that run into the 2040s and 2050s. What remains unspoken is the heavy question, how far out into the future does it even make sense to plan? I look at the figures projecting our presumed spending when my wife and I will be old, we hope, and our son in the prime of his life, we hope, and the numbers seem to wiggle and shimmer into abstraction, shedding whatever meaning they hold today and reverting back to mute patterns of lines and curves. What meaning *can* they possibly retain when we know that by then the world will be shaken by so many unknowns both known and unknown? The magnitude of the disjuncture between what we know and what we experience on a day-to-day basis embodies the degree of the dissonance we feel, and the weirdness of the juxtaposition itself accentuates the feeling. It's a vicious circle that easily spins us off onto other off-ramps, from apocalyptic fatalism to morose despair.

What this disjunct is, is a sense of what Amitav Ghosh and other writers have termed "the uncanny," the feeling that even as life goes on as it always has, it has shifted in an irrevocable way, grown shaky on its axis.[25] This is ordinary life; this is not ordinary life. We find ourselves in a horror film, early on, when even as some characters have already experienced the worst, others still hang on to the hopeless hope that things will not go too awry. If a visceral sense is what's required to really take a threat or opportunity seriously, climate breakdown, for many, remains in the uneasy no-man's-land between intellect and gut, subject to constant negotiation, recalibration, and doubt about how much credence to give it. It is real; it isn't real; it dominates my thoughts; it can be disregarded for a while longer. All these ways of seeing seem true, but because they contradict one another all of them are inevitably in question too. "The possibility of climate change is both deeply disturbing and almost completely submerged, simultaneously unimaginable and common knowledge," the sociologist Kari Marie Norgaard writes in her book *Living in Denial*. "People care and have considerable information, but . . . they don't really want to know and in some sense don't know *how* to know."[26]

In the case of climate breakdown, cognitive dissonance arises in part from a confusion between two different ways of experiencing time—something the world also gained some experience of during the coronavirus epidemic, when the exponential progression of infection and the inertia of everyday life clashed in so many ways. People, and societies, inevitably experience shocks and moments of extreme change. But people and societies alike don't survive unless they also go through extensive patches of quotidian time—the day to day. Refugees fleeing for their lives still have to sleep along the way; even the most well-equipped army goes nowhere without being fed. The ancient Greeks dealt with this paradox by explicitly naming what they conceived of as two different categories of time. *Chronos* was ordinary time, the measure we use to assess how long to boil an egg, how long it will take to get to work, what the week's schedule looks like. It is what we use to mete out our lives. It is the normal. It is the predominant way in which we experience the unstoppable and irreversible passage of hours and weeks and years. At a societal level, it consists of feeding the army, or growing the food. Chronos is distinguished by the quality that the fundamentals of life don't change as it passes.

The other kind of time they called *kairos*, which refers to events that represent a profound break with the past. The army does its work: war is declared, won, lost. An earthquake demolishes a town. A democracy is declared, or taken over by a dictatorship. For the first time, a single bomb levels an entire city. A moment of kairos changes everything. The order of things ruptures. The world turns upside down.

At a personal level, but often not a global one, adults grow accustomed to the knowledge that kairos always threatens the smooth chronological flow of our lives, for good or ill. In an unexpected moment you might meet a true love, or be crippled in a car crash. You might find out you're pregnant or be diagnosed with cancer. The ultimate kairos events—being born and dying—also represent watershed moments for those close to you. Looking back, we readily identify the kairos moments of fate and decision as the core skeletal structure of our lives, the beams and studs to which the chronos of everyday life is fastened.

The stories of our lives consist of the rhythmic interplay between these two types of time, between tension and relaxation. The juxtaposition of chronos and kairos is always curious, always fraught, always worth remembering and examining; no wonder it fuels so much storytelling.[27] In how many unexpected moments has a Hollywood hero encountered a true love, or a nemesis, in what initially seemed the most mundane of settings? The narratives that glue together a society are made up of the back-and-forth between kairos and chronos moments, between ebbs and flows of tension, just as our individual and family lives are.

What the ancient Greeks, as well as myriad other peoples throughout the world, knew was that both types of time are critical to societal well-being. Days and weeks and years are made of ordinary chronos moments in their ceaseless flow, but because kairos moments inevitably come too, societies have to develop ways to deal with them. A society that cannot grows brittle and shatters when it is stressed. Our society faces the opposite threat: it has been acting as if it finds chronos boring and requires regular infusions of kairos to make life seem worth living. We are living in what the French philosopher Guy Debord in 1967 termed a "society of the spectacle," which he defined as a society in which economic pressures that had previously shifted life's emphasis from *being* to *having* now emphasize *appearance* above everything else.[28] He defined the "spectacle" as "a social relationship between people that is mediated by images. . . . It asserts that all human life, which is to say all social life, is mere appearance."[29] And a story comprised of images is boring unless each new image exceeds the last in excitement. We want each of our moments, too, to surpass the one before. For a long time the United States has collectively acted according to the belief that stasis is unacceptable, that the unchanging is deeply boring, that chronos is inadequate unless it incorporates an element of kairos.

Climate change, Robert Jay Lifton has written, explicitly scrambles our understandings of these two types of time. "We have generally associated climate with *chronos*, but with increasingly frequent and greater catastrophes it veers toward *kairos*," he writes. "Climate normality covers over a threat that hovers between *chronos* and *kairos*, confusing our relationship to time and our ways of imagining the future. Holding on to climate normality becomes ever more difficult."[30] Chronos explicitly disappears from everyday life for those who experience the immediate impacts of climate change—the wildfire, the flood, the hurricane. But it also continuously threatens to disappear for all the rest of us who may not be experiencing the immediate impacts but who get swamped by the knowledge of what is happening elsewhere and what is likely to happen on a much broader scale before too long. The foreknowledge of breakdown to come takes away the contemporary comforts of chronos. It has become a cliché to speak or write of some novel environmental condition as the "new normal." More apt, when it comes to climate breakdown, is the fear that there will simply be no such thing as normal anymore.

Most of the world had a foretaste of this dissonance during the coronavirus pandemic, which ripped away many of the certitudes of chronos and replaced them with profound uncertainty about what was to come in the realms of policy, economics, work, social life, and personal life. Like climate breakdown, the pandemic quickly transitioned from being an "issue" to being a hyperobject, an all-encompassing context for new experience, conflict, anxiety, and decision-making that could scarcely be avoided. Weekly, daily, sometimes even hourly, residents of the United States and other wealthy nations who previously might have felt well buffered from many "global" problems found themselves facing new existential questions about risk and survival, often in a setting in which the reliability of once solid information was repeatedly called into question. It was like finding oneself stranded on the melting Arctic ice pack, frantically jumping from one floe to another in search of solid ground. And as my colleagues Diana Stuart, Brian Petersen, and Ryan Gunderson have pointed out, the pandemic was subject to some of the same sorts of self-interested false story lines—notably denial, an undue focus on individualism, and techno-optimism—that have also characterized the climate crisis.[31]

Scholars in the environmental humanities use the word *storyworld* to refer to the created world of a narrative, a place into which we can transport ourselves through the act of imaginative transposition. Middle-earth is a storyworld; so is the Hogwarts School of Witchcraft and Wizardry;

so are Dickens's depictions of Victorian London and Toni Morrison's nineteenth-century Ohio in *Beloved*. "In addition to providing readers access to new, possibly unfamiliar environments," the ecohumanist Erin James writes, "storyworlds also provide readers access to highly subjective understandings of what it is like to live in, conceptualize, and experience a given space and time."[32] Playing on concepts of *like* and *unlike* as metaphors do, a fictional storyworld highlights both its similarity to and its dissimilarity from the time and place in which readers find themselves. Constructing one, James explains, requires of readers active work of the imagination, "an inherently comparative process, in which readers come to recognize the subtle and not-so-subtle differences between the world their physical bodies occupy and the world to which their mental energies are directed to by the reading process."[33]

In an era when scientific understanding of a climate breakdown future is getting better and better, our shared apprehension of how we will be living soon constitutes a sort of storyworld, a collective glimpse of fire, flood, drought, deprivation, anxiety. The climate breakdown storyworld is a science fiction narrative glimpsed over and over again in the act of springing to life, or transforming from fiction to nonfiction, with its recognizable components of our lives today slid askew by the new pressures of nature and society both gone haywire. It is the world of today seen in a not-so-funhouse mirror. It is uncanny enough to think of the highly plausible physical manifestations of climate breakdown becoming part of our lives: tidewaters in the streets, killer heat waves, droughts in agricultural breadbaskets, storms of previously unknown ferocity. It is worse still to think of how societal responses to these physical effects will ripple and multiply through our communities.

Here's what my climate breakdown storyworld looks like to me as I write. I began writing this book during a year—2019—in which the Southwest's summer monsoon largely went missing, leading to a miserably dry summer characterized by ongoing wildfire danger and drought-stressed forests. I continued writing throughout 2020, that pandemic-wracked year. Isolating at home as so many others did, I'd never had so much time to spend in my backyard garden. Greens, beans, corn, squash: I babied them through the always hot and dry weeks of June, watering in the cool of morning, longing for the sound of thunder on the horizon that often begins around the Fourth of July. *Last year's monsoon was dry*, I calculated; *surely chances are this one will be good, right?* Weather may be a crapshoot, but over time the extremes have always balanced out.

But no. July, the typical beginning of the monsoon season, arrived with a couple of scattered storms, but then the clouds withdrew to the south again. Consistent high pressure kept the moisture from working its way north from Mexico. Late that month a good soaking was forecast, days of storms. But no: after a single drenching day clear skies took hold again.

A year earlier, I'd felt a high degree of confidence that a missing monsoon season had little to do with climate change. Now the second dry summer in a row instilled both a deep sense of foreboding and troubling midnight thoughts that made no rational sense. *Maybe the wall is working*, I found myself thinking, though I knew that the steel barrier the Trump administration was building along my state's southern border couldn't possibly be high enough to affect the movement north of subtropical moisture. But that rational understanding hardly mattered then, or now. Just as I used to dream of Middle-earth when as an adolescent I binge-read *The Lord of the Rings*, I continue to find myself relating to the climate breakdown storyworld in a dreamlike way. Is it something that belongs to the future, or has it protruded into the present? I often can't be sure. When I sit in my backyard writing shack, typing away, the world around me largely makes sense, while the future world of climate breakdown with its unpredictable varieties of chaos does not. But when I enter thoroughly enough into picturing that chaos it is the (mainly) ordered world of today that seems to make no sense. It's like the weird juxtaposition that takes place when you awake from a vivid dream. You've been in a setting that has an internal logic to it, but when you wake you jump swiftly into another setting with its own deeply felt structure, and the logic of the dream instantly seems bizarre, incomprehensible. In the bright light of morning I know that the new border wall can't have anything to do with the missing monsoon, but at night, I know, I'll soon be making such conflations of dread again. And so I fear the next coming monsoon season as much for the meaning I might find in it as for the actual physical presence or absence of water and moisture and clouds themselves.

Psychology provides us a word for this disconnect. In 1987 the psychologist Paul Slovic summarized a number of studies looking at how people perceive risks in an influential article that appeared in the journal *Science*.[34] How people feel about risk, he concluded, depends on the interaction between two factors. One is how "known" a given risk is—which includes such factors as whether it's new or not, whether scientists have explained it or not, and whether it's directly observable or not. Auto accidents, for example, were very well known in the 1980s, while threats associated with DNA technology were not. The other factor is what he labeled "dread," which relates to the consequences of choosing particular courses of action, especially the extent to which those consequences are controllable and limitable. Even though backyard swimming pools kill and injure a significant number of people, especially toddlers, each year, they were ranked very low in dread because people who have them feel they can control their exposure to the risk they embody. The prospect of nuclear war, which topped Slovic's "dread" scale, instills great fear precisely because it seems entirely uncontrollable to everyone except perhaps a handful of elite decision-makers, and because we believe that such a war would cause practically limitless death and destruction.

Climate change did not appear on Slovic's list of risks because it was not yet a thing back then to psychologists. But it is now, and it is clear that the prospect of climate breakdown has come to occupy a very high position on the dread scale. Like nuclear war, it threatens not only those living today but the quality of life and even the possibility of life for future generations. Its consequences seem limitless both in space and in time; unless you're Elon Musk planning to decamp to Mars, there will be no escape from it. Those consequences cannot be reconciled with most of the standard tools we use for making decisions in the public sphere. Nor can we square them with the sorts of decision-making we engage in every day in our chronos lives, or with the feel of daily life.

It is this deep sense of dread, then, that can easily become a barrier to action for those who in no way are climate change deniers or end-times investors. "Is it possible to be haunted by the future as well as the past?," the essayist Emily Raboteau asked about her need to think about climate change in a post-Sandy New York City.[35] The answer, of course, is yes. And when that haunting reaches the level of dread, it can easily become paralyzing. Dread equates to seeing no way out, no plausible course of action. Dread is what whispers in my gut that if I were in the horror film, with the knife-wielding maniac lunging out from the darkness of the basement stairs, I could do no more than stand helpless, too dumbstruck even to scream.

Perceiving our society's widespread sense of climate breakdown as instilling dread can help us to better understand the various kinds of denial with which we approach the problem, and the interconnections between them. The sort of blatant antiscience denial epitomized by right-wingers is an effort to slide climate change back down the dread scale, to dial its potentially limitless effects down to finite and manageable problems of building carbon-sucking infrastructure and higher seawalls, to make it an "engineering problem." But dread is a terrible enough emotion that those who are upfront about the genuine fear they feel have ample reason to pull back from it too.

There are numerous psychological mechanisms for this. Climate breakdown is a powerful threat to the "just world bias," which holds that the arc of the universe bends toward light rather than darkness—a powerful belief in a historically optimistic nation like the United States.[36] More universal still is the human tendency toward what psychologists have termed "terror management," which is above all centered on the knowledge that we or what we care for will die. A fear of death, writes the Finnish eco-anxiety scholar Panu Pihkala, "very often causes people to strongly defend the current status quo of things as a defense mechanism against uncertainty."[37] The more we know about the climate breakdown storyworld, the more firmly many cling to the ultimately illusory securities of the present day, including, perversely enough, more shopping.[38] And even a full embrace of the dismal predictions of climate breakdown can lead to paralysis owing to what psychologists have termed "pseudoinefficacy."[39] This phenomenon occurs when people realize that they can by acting address only a small part of a large problem; in the face of one's incapacity to fix the big things, it is easy to not even try to fix the small things. The bottomless nature of the climate breakdown future, the story line that things will only get worse as physical and social feedback loops further feed one another, serves as a perfect invitation to look away.

"Many people in fact care too *much*, not too little, and as a result they resort to psychological and social defenses," Pihkala writes. "Full-scale denial is only one of those responses. Different forms of disavowal are much more common. People find ways to both know and not to know at the same time. This results in a vicious circle. Because of denial and disavowal, the problems get worse. This in turn breeds more anxiety, which many try to deal with by more denial, and emotional pressure keeps building."[40]

Taken together, these psychological mechanisms, all of which in their proper setting can play a valuable role in cushioning us against traumatic psychic blows, can easily serve as barriers to action even for those who have few illusions about climate breakdown. Per Espen Stoknes, the Norwegian psychologist who wrote the book *What We Think About When We Try Not to Think About Climate Change*, describes the frequent consequence of climate breakdown dread as the "5 Ds": distance, doom, dissonance, denial, and (i)dentity.[41] What he means is that we work to conceive of climate breakdown as something far removed from us; we resist knowledge of how bad it may be; we feel a strong disconnect between our daily lives and what we know of the future; we as a result look for narrative off-ramps that will comfort us with a somewhat plausible alternative; and we look to reinforcing tribal identities that will support our choice of that narrative. What's important is that virtually all of us do this, regardless of political affiliation or belief. We are bound by dread, and by our common responses to it, even if those responses pull us in very different directions when it comes to policy, behavior, and how we identify ourselves.

This can be true even for extremely well-versed scientists. A growing strain of evidence has documented how a long series of seemingly authoritative climate change reports has tended to understate risks. "Climate skeptics and deniers have often accused scientists of exaggerating the threat of climate change, but the evidence shows that not only have they not exaggerated, they have underestimated," the science historian Naomi Oreskes and her co-authors wrote in 2019. "Scientists tend to underestimate the severity of threats and the rapidity with which they might unfold."[42] In particular, analyses have concluded, such phenomena as the thawing of permafrost, releases of methane from the Arctic, and the melting of ice sheets and the resulting sea-level rise seem to be progressing faster than the most widely accepted models have predicted.[43] This happens for several reasons. One is social: projections based on consensus tend to slide toward a sort of watered-down middle ground in which conclusions that appear more extreme are rejected or downplayed. That's especially true for reports sponsored by governments, such as those from the IPCC. But Oreskes and her co-authors argue that individual scientists too dilute their climate science stories by generally hewing to more conservative projections of what might happen.[44] And others simply tend to leave out difficult-to-measure components that might otherwise go into climate models. "If we can't quantify something very well, we tend to ignore it," is how one climate modeler explained this tendency to the journalist Jonathan Mingle.[45]

These tendencies toward caution have had the effect of cloaking the more devastating potential consequences of climate breakdown. An increasing number of observers argue that it is vital for humanity to more seriously consider the consequences of so-called "fat tail" climate scenarios, meaning those that lie on the further reaches of probability but that would carry the most severe consequences—including, according to some, the likely vanishing of human civilization as a result of chronic and multiplying stressors.[46] "The failure of both the research community and the policymaking apparatus to consider, advocate and/or adopt an existential risk-management approach is itself a failure of imagination with catastrophic consequences," write the Australian analysts David Spratt and Ian Dunlop, likening humanity's behavior to that of an airline passenger who knows that the plane has a 5 percent chance of crashing.[47] Would you choose to board? For better or worse, almost all of us continue to stand on the jetway, eagerly awaiting the flight, even as the fine print on the boarding pass recites ever more dire warnings.

These failures of analysis are much more the fault of science as a public service or means of investigation than of individual scientists themselves. But they also reflect that scientists are human beings too, subject to the same range of emotional rewards and pitfalls as everyone else. Even as climate scientists in particular make observations and construct future scenarios that show that things may get very bad indeed, they undergo the same imperatives to get through daily life and to protect themselves emotionally as the rest of us. Even as they dig deeply into their science, they are apt to reach for the same tools of disengagement as everyone else. They keep "the heart a long way from the brain," as the geographers Lesley Head and Theresa Harada wrote about climate scientists they interviewed, utilizing "a range of behaviours and strategies to manage their emotions around climate change and the future. These include emphasizing dispassion, suppressing painful emotions, using humour and switching off from work. Emotional denial or suppression of the consequences of climate change worked to enable the scientists to persevere in their work."[48]

It is no wonder that in the face of all these pressures, and of all these tempting alternative ways of thinking or feeling about climate breakdown, one of the strains of thinking that has grown most rapidly in recent years is the apocalyptic, the ultimately Romantic idea that there is no way out. There is nothing to be done, according to this narrative, except to ensure that humanity finds as much grace as possible as our future prospects collapse along with the climate. It's a genre epitomized by David Wallace-Wells's 2019 book *The Uninhabitable Earth*, or at least by the media reception of his book.[49] But there are many other examples out there, such as the "Deep Adaptation" paper published online by the British academic Jem Bendell in 2018, which had the stated goal of providing "readers with an opportunity to reassess their work and life in the face of an inevitable near term social collapse due to climate change."[50] It has been downloaded hundreds of thousands of times and has spawned an online network of fans.

If the range of forms of denial constitutes an ideological spectrum, as in the seating of members of a parliament, the narrative of Romantic fatalism practically makes of the spectrum a full circle, for some of its adherents find as little reason to do anything about the impending breakdown as those who continue to deny there's any problem at all. For both, the pursuit of a story that appears to adequately explain what's going on outweighs any urge to do the sort of often messy planning for the future that we readily embrace in other realms of life.

For most of our history as a species, the creation of meaning—of a life's narrative—was not a matter fraught with existential angst. It was something of a given. This is not to say that there were not difficult choices to make, or that living was easy, or simply a paint-by-numbers matter of living out the dictates of God, or Fate. But it is to say that patterns of meaning—in spiritual belief, in relations to a community's environmental surroundings, in the workings of an economy—were largely established in advance, through long tradition and mutual consent, and codified in a narrative in which society's members all held a stake. There were rules, and there were reasons for the rules, and an agreed-upon narrative, far exceeding the confines of individual lives or generations, tied them together in a coherent way.

"Myth," wrote Debord, "was the unified mental construct whose job it was to make sure that the whole cosmic order confirmed the order that this society had in fact already set up within its own frontiers."[51]

This fundamentally conservative way of organizing human life fell apart in several iterations. The first was the appearance, in the ancient Middle East, of monotheistic religions that substituted linear narratives of history for the older, generally circular conceptions of time that prevailed in both hunting-gathering and agricultural societies. The result was the development of what Debord called "irreversible time," in which history takes the form of an inevitable progression toward a defined end goal—in the case of Christianity, the Last Judgment with its accompanying denouement of Heaven or Hell.[52]

The narrative arc of tension, codified in writing by the Greeks for purposes of entertainment and education, now had a pronounced purpose, a divine mandate. No wonder that when technological innovation, new ways of thinking about individualism, and geographic good fortune allowed the West to implement expansionist ideas about capitalism and colonialism they dovetailed conveniently with what was already a core religious belief. Now the arc of history that already bent toward a higher, ultimate purpose could simultaneously accommodate and encourage endless expressions of individual, corporate, and national ambition. The result was societies in which many of the old rules did not have to apply. To some extent in Europe, and to a greater extent in the new (for Western civilization) lands of the Americas and Oceania, settlers and residents developed societal and economic systems in which the making of a life's narrative could take place across a much wider field of play. In these new societies, you could reinvent yourself. You could light out for the territories, walk away from a troubled past, forge a new identity, sniff out opportunities that would have been closed off in a more hidebound place. In New York, it turned out, you really could be a new man.

But the one thing you could not change, because it was too integral, a sort of citizenship requirement for those who would succeed in these new nations, was the obdurate belief in improvement and growth. The job of every citizen, of every societal grouping, of the nation itself, it lined up perfectly with core religious foundations and with our constant psychological readiness for stories that embodied it. It was the diamond-hard foundation that led to narrative expressions of America as the shining city on a hill; as a frontier that pushed west, then overseas, then into space, even as the leading edge of the moral arc of the universe. And as a deeply shared belief, it is the reason why politicians and governments on both sides of the political aisle have been almost equally toothless in combating climate change. As the Australian writer Clive Hamilton has phrased it, "The pressures on each of us to immerse ourselves in the stream of Progress and be swept along by it are almost irresistible. Our freedom then becomes only the liberty to navigate to the left or right side of the current, or, if too weak, sink to the bottom, there to drown."[53]

But history is constructed by those who swim in the mainstream, and the result of their work has been societies in which the creation of meaningful life narratives became a much more important and individualized preoccupation than had ever been the case before. It is no coincidence that Hollywood sprouted in California, at the far western edge of the US frontierlands, in a sun-kissed part of the country that was an especially strong magnet for those seeking to reinvent themselves—and no coincidence either that it was this dream factory that came to dominate the collective imagination of the world throughout the course of the twentieth century. If society provides the freedom to imagine one's own life story, it also offloads the responsibility for creating those stories onto individuals, and with the weight of that responsibility on our shoulders we look for guidance to the most compelling models we can find. And so for a century now Hollywood, with its cousins Madison Avenue and Silicon Valley, has been producing an embarrassment of riches whose abundance provides innumerable and indispensable guides to crafting our life stories.

Curating has become the contemporary term of art for that work, and to curate one's experiences so that they collectively make sense has become the vital task of modern citizens, a daily or for many hourly ritual that often involves the sharing of practically every experience through social media and the constant tabulation of how it registers with others—indeed, the *structuring* of lived experience so that it makes sense on social media. We have become so accustomed to shaping the raw material of our lives into a sharable narrative, in real time, that it has become exceedingly difficult to draw a distinction between an experience as a time-bound manifestation of raw and unique individuality—no one else has ever experienced exactly this combination of place, relationship, conversation, emotion, thinking before—and as an element of narrative that we can use to organize reality in a relatable way. We have come to line experience up in service of our preconceived narrative. "The American way of life is consciously about language, storytelling, plot and form, and is meant to draw attention to its own status as fiction," writes Bruno Maçães in his recent book, *History Has Begun*. "Rather than something that emerges from experience, communication becomes the very point of experience and its organizing principle."[54] And as we communicate our selves, many of us lean heavily in the direction of that time-tested American optimism, limning a generally sunny or at least coherent portrait of how our lives are going.

We draw lessons in both content and structure from fictional stories, then, and apply them to the real world around us. (Here again, politicians were ahead of the academics, as the rise of Ronald Reagan proved in 1980.) "Truth is hard work," the artificial intelligence expert Elizier Yudkowsky has written, "and not the kind of hard work done by storytellers. We should avoid not only being *duped* by fiction—failing to expend the mental effort necessary to 'unbelieve' it—but also being *contaminated* by fiction, letting it anchor our judgements. And we should be aware that we are not always aware of this contamination."[55] As the world grows more chaotic as a result of climate breakdown and other instabilities, many grow increasingly desperate to see real life align itself with the sort of consistency that we find so appealing in fictional stories. The more it looks as though the smooth predictability of the climate is going to be taken from us, the more we are primed to find some satisfying story line that makes sense to us. Conspiracy thinking, which in the case of climate change is much more rampant in the United States than in most other Western countries, is a quest for a sensible narrative that arises when more mainstream, and generally more rational, explanations are simply not palatable. Indeed, a number of studies have shown that those who are more likely than most to engage in conspiracy thinking are much less likely to admit that climate change is a problem.[56]

As the *New Yorker* television critic Emily Nussbaum has written, story lines in the Republican Party, especially, have "drifted further and further from any basis in factual reality. . . . They have been forced to become the party of Scheherazade, telling a series of mesmerizing stories in order to stay alive."[57] The result is the media-political dystopia that culminated (so far) in the election of Donald Trump, and that the novelist and media critic Kurt Andersen has labeled "Fantasyland":

> Each of the small fantasies and simulations we insert into our lives is harmless enough, replacing a small piece of the authentic but mundane here, another over there. The world looks a little bit more like a movie set and seems a little more exciting and glamorous, like Hitchcock's definition of drama—life with the dull bits cut out. Each of us can feel like a sexier hero in a cooler story, younger than we actually are if we're old or older if we're young. Over time the patches of unreality take up more and more space in our lives. Eventually the whole lawn becomes AstroTurf. We stop registering the differences between simulated and authentic, real and unreal.[58]

This, then, is the frame of mind that we bring to bear on our plight. Addicted to narratives with a sensible and satisfactory progression, we slip easily into perceptions of climate breakdown as a simple matter of revelation or a problem to be solved or cause for despair—and so time and again we turn away. Caught up in the unholy melding of experience, politics, entertainment, and definitions of self that has come to dominate our attention and our polity, we are unable to grasp the fullness of a problem that has surpassed the bounds of Fantasyland and become, simply, Realityland. At times it seems that the only things that unite us anymore are twofold: our addiction to the spectacle that we are collectively living today and our dread of the future that we know is coming. Both are expressions of the yawning gulf we feel between what we know and what we suppose we have the power to do. And so even as we feel worse and worse about our prospects, we face the real possibility of doing little more than continuing to act as spectators in a reality show starring all of us.

ADDRESSING THE DISQUIET ON THE LAND

"One of the penalties of an ecological education," wrote the wildlife biologist Aldo Leopold in the 1940s, "is that one lives alone in a world of wounds."[59] He was writing about such examples of human mismanagement as the extirpation of predators, the loss of fertile topsoil, and degradation of ecological communities, and about what seemed at the time the rarified sensitivity needed to perceive those alterations. Today the wounds are more severe, with worse to come. But alone? No. No one need be alone anymore in coming to terms with climate breakdown. The signs are too clear, the understanding of what's to come too widespread.

That doesn't mean that many don't *feel* alone. Surveys have shown that even as the great majority of Americans—and people elsewhere—know about climate change, a surprising number hardly talk about it.[60] In too many public settings the topic is still fraught with treacherous implications or assumptions about political identity. Even where there is little dispute about the existence of, causes of, or seriousness of climate change, it is still often avoided in conversation because it brings up too many difficult issues of powerlessness, complicity, anger, and loss.

The result is an unsalved emotional wound that itself contributes to feelings of paralysis and an inability to address the problem head on. In failing to acknowledge environmental grief, writes the educator Elin Kelsey, "it's as if we are engaged in a mass movement of emotional denial"—one we can turn around only by being far more explicit about what the emotional costs are.[61]

To accomplish this, educators and therapists are taking a leaf from the practice of psychology by encouraging people to acknowledge and address climate grief.[62] If we can move past the death of a loved one only by openly coming to terms with the grief of their absence, the thinking goes, we can only be honest about climate breakdown by first naming and processing the grief we feel. That's true whether the emotion wells up from the experience of what the Australian philosopher Glenn Albrecht terms "solastalgia," by which he means the pain people feel as a beloved landscape is degraded, or from the premonitory feeling of anticipatory grief that comes of knowing something about future climate change projections.[63] The sense of loss we feel, writes the Buddhist scholar Joanna Macy, is not something to be avoided or feared but rather embraced. "Like living cells in a larger body, it is natural that we feel the trauma of our world," she and her colleague Sam Mowe wrote in a recent essay. "So don't be afraid of the anguish you feel, or the anger or fear, because these responses arise from the depth of your caring and the truth of your interconnectedness with all beings."[64]

The work of understanding this connection has resulted in an upsurge of interest among mental health professionals who have seen their patients grappling with the emotional toll of climate breakdown.[65] The American Psychological Association proffered a clinical description of "eco-anxiety" in 2017 as professional analysts began to see rising rates of anxiety, depression, and even suicide attempts tied to their patients' worries about climate change. A new Climate Psychiatry Alliance was founded that same year.

Some of that mental health work is for professionals, but advocates say that the task is large enough to warrant much broader participation. Some have explicitly turned to the arts, which in their ability to move beyond words have long been sources of solace and inspiration as people have faced challenging times. The philosopher Kathleen Dean Moore teamed up with the pianist Rachel McCabe to create a performance titled *A Call to Life: Variations on a Theme of Extinction,* which paired Moore's reflections on impending loss in the natural world with McCabe's interpretation of Sergei Rachmaninoff's somber Variations on a Theme from Corelli.

But perhaps the most telling evidence that people recognize the importance of dealing with climate emotions lies in the rapid expansion of the Good Grief Network, founded in 2016 by Aimee Lewis-Reau, then a University of Utah graduate student.[66] It explicitly addresses climate change–related grief by providing a process for converting feelings of loss and sadness into spurs for action. Roughly based on the Alcoholics Anonymous model, the program centers on a series of meetings that take participants through a sequence of steps designed to put them in touch with their feelings, acknowledge complicity, face mortality, look for beauty even in a damaged world, and reinvest energy in positive action.

Participants, Lewis-Reau told a journalist in 2019, experience "a shift in energy, in body language, and enthusiasm for life. They come in defeated and anxious, shoulders high, they leave more present, more grounded, with clarity."[67]

Unsurprisingly, participants in the network's rapidly growing number of local groups are largely confined to those who have openly identified climate change as a major problem. Among them, there is plenty of room for growth in the alchemy of converting grief into action. But emotional denial extends much farther. Insofar as more than 70 million Americans voted for an outright climate change denier in 2020, the larger task may be to extend the processing of climate grief to those who can't yet admit they need it.

4

TRAGEDY

Of all modern myths, the one our culture clings to most tenaciously is the myth of human control. Perhaps it would even prefer oblivion to letting this myth go.
—Paul O'Connor[1]

It should be clear by now that when I so repeatedly refer to the "stories" or "narratives" that societies turn to in order to define themselves, what I am really referring to is something akin to *myth*—that is, a type of story to which we grant unusual power, and whose foundations are sufficiently hidden that we fool ourselves into believing that it is not a story at all but rather an unquestionable truth. We need myth to provide meaning to our lives at both an individual and a societal level. This is as true now as it has ever been, even as many contemporary people tend to view mythic stories as hoary relics of defunct worldviews. It is, after all, a fundamental characteristic of myth that the myths providing meaning to other cultures often appear ridiculous, even as those of one's own seem self-evident enough to fade into invisibility.

In the previous chapter I explored how many of the foundational myths of America have coalesced into a broadly shared self-identity that poses enormous hurdles to dealing with climate breakdown. But even though those myths are central to the nation's self-identity it is plausible to understand the connection between them and climate breakdown as fundamentally coincidental: it may be deeply unfortunate that we came to believe those particular things about our nation and ourselves because those beliefs predispose us to have a really hard time grasping and dealing with the consequences of climate change, but the process has not been deliberate. That we have so thoroughly failed in our task of understanding the breakdown before us might be no more than an accident of history. Such an analysis would have the benefit, from the standpoint of assigning guilt, that our failures are not really our fault. They are, instead, a side effect, an unfortunate consequence of decisions collectively made long before climate change swelled into a hyperobject to worry about.

But there is another and more disturbing way of looking at the connection between the myths we live by and climate breakdown, and that is to understand some of those myths as amounting to tragic flaws in the makeup of our society.

I use *tragic* here in the sense that comes to us from distant roots in ancient Greek drama, where *tragedy* was understood as an unavoidable narrative arc, its progression merely the physical manifestation of all that has been present from the beginning. For those living in tragedy, nothing can change the outcome of the story. If you're Oedipus, it is inevitable that you are going to kill your father and marry your mother; no amount of self-awareness or determination can change that trajectory. Antigone is going to bury her brother even though the act is politically proscribed and will bring on her own death and that of her lover. "Tragic figures bring on their own suffering," wrote the literature scholar Joseph Meeker in his 1974 book *The Comedy of Survival*, "for they have taken a course of action which must inevitably lead to their doom, even though they may not have been aware at an early stage of the consequences of their choice."[2] Tragedy occurs when character flaws or life trajectories are readily evident but cannot be overcome. It's fate, and fate is decreed by suprahuman forces, whether it's the literal Fates who predetermine the duration of human life spans for reasons that seem random to us or gods who intervene out of lust, spite, pettiness, or other typically all-too-human emotions.

What I am getting at here is the sad but vital realization that our adherence to myths that make it nearly impossible to deal with climate breakdown is not circumstantial but rather tragic. Tragic, because even though the myths themselves predate widespread understanding of the global atmosphere and how it can be affected by our actions, our continued adherence to them in the face of what we have known for some time represents an embrace of inevitability that suggests we are as helpless to alter our course as Oedipus was his. Even as we have known that our actions—or inactions—will have the consequence of making life immeasurably worse for our children, or grandchildren, or other generations who may come beyond them, we have stayed on the same track, using our considerable abilities of self-knowledge primarily to double down on our destructive behavior.

In Greek drama, the plot was generally well known to audiences before the play began because characters were drawn from traditional mythic stories. What audiences wanted to see in the grand stone amphitheaters was not surprise twists in the action but rather how character came to be revealed by how gracefully or petulantly the heroes or antiheroes accept their situations. Ever since I read my first Sophocles in high school I've wondered if some actor did not feel the temptation to improvise, to surprise, to diverge from the script rather than play along. What a revelation that could have been to the audience: *You can do that? You can change what seemed foreordained?* But no: Oedipus was doomed to follow his bitter path before he ever showed up on stage. In our society's adherence to certain foundational myths we appear to be following the same path, running headlong toward a tragic outcome even as its outlines grow ever clearer. And there is one myth in particular that most embodies this tragic worldview, namely, that the nation exists primarily as a foundation—a stage set, if you will—for a particular kind of economic activity, and that this role supersedes others that a society might imagine for itself.

This myth is tragic because it posits that the primacy of "the economy" is inevitable regardless of the consequences. And if it is inevitable, then we simply lack the ability to do anything but continue to live according to these foundational myths until the end comes. "There is tragic irony," Meeker wrote, "in the fact that man has achieved the long-sought mastery over nature only to find that his very existence depends upon the natural balances which were destroyed in the process."[3] To conceive of climate breakdown as a tragedy rather than as some other sort of story is to concede that we can do nothing about it—and this is the story that has been hammered home, explicitly or tacitly, by many of the most powerful people and institutions in the world. However powerful humans may be on the world's stage, they suggest, the script we follow is written not in pencil, nor even in pen, but in hard stone.

The economist Kate Raworth points out in her book *Doughnut Economics* that the conventional economic models that have since World War II dominated and shaped decision-making about policy—not just in the United States, and not just in Western democracies, but increasingly all around the globe—have a major flaw.[4] It's that they reflect the static understanding of "nature" and of "climate" that we examined in chapter 1, namely, that these fundaments of all life on Earth provide a stable and predictable backdrop against which human life and economy are able to play out. In the case of the standard economic models that remain an elementary building block of textbooks, Raworth writes, this assumption posits that nature is both an endless source of raw materials and a bottomless sink for the waste produced by human economic activity. Climate, meanwhile, is never mentioned at all—it is assumed to be unchanging, an endlessly reliable and silent partner that makes it all possible for us. Mother Earth, in this telling, is the classic female partner of misogynists' dreams, a fecund helpmeet who is always there to assist and never raises objection.

Raworth is far from the first economist to identify such shortcomings. The 1972 report *The Limits of Growth* famously critiqued conventional economic models' growth-at-all-costs paradigm, pointing out that both resources and the planet's ability to absorb pollution are far from unlimited. Also in the 1970s the ecological economist Herman Daly proposed the "steady-state economy" as an alternative to the mainstream growth model. The human economy, he noted, cannot exist without inputs of energy and raw materials, either of which can place absolute restrictions on how much humans can do with them; "in a finite world," he noted, "continual growth is impossible. . . . The earth approximates a steady-state open system, as do organisms."[5] Human societies could only become sustainable, he suggested, if they could rein in their appetites. These critiques constituted a strong indictment of mainstream economics, if only because the points they make are so obvious that they point to a severe myopia among economists. But what is particularly intriguing about Raworth's recent critique—*Doughnut Economics* was published in 2017—is how thoroughly she couches it in terms of story. She explores how economic models are often portrayed in graphic terms, such as the "circular flow" diagram first developed by Paul Samuelson in the post-war years, which lays out how businesses, households, banks, government, and trade interact to move goods, services, and money among one another. Missing, of course, is any mention of climate. It is an economist's drawing of the stable stage set for human affairs: in failing to assign climate a role, the diagram implicitly suggests that it cannot change and hence does not need to be considered in decision-making.

Fine, you say, but that was 1948, when only a handful of scientists had any inkling of the global changes our greenhouse gas emissions were capable of initiating. But circular flow diagrams also don't include inputs from nature, either in the form of raw materials—soils, biodiversity, metal ores, limited groundwater supplies—or in the form of energy, and such diagrams have continued to serve as a handy shorthand for how the economy works well into the twenty-first century, long since ecological verities have become better known. And the diagrams themselves long ago went viral. Raworth points out that because they communicate so much in such a terse, memorable form, they are extremely potent summations that carry a much bigger punch than lengthy verbal explanations. That's as true for the circular flow diagram as it is for the chart of skyrocketing GDP. Such graphic images "linger," Raworth writes, "like graffiti on the mind, long after the words have faded; they become stowaway intellectual baggage, lodged in your visual cortex without you even realising it is there. And—just like graffiti—it is very hard to remove."[6]

No wonder generations of economists came to describe factors that don't fit into the diagram—such as the carbon pollution that's changing the climate—as "externalities" that simply have no place in economics equations, or in investment or policy decisions. Because waste didn't have a place in the diagram, there was no reason to factor it in as a cost, or as a resource for someone else. Even as it becomes more and more clear that our relentless economic activity amounts to taking out pieces of the stage set on which our societies are standing, leaving the entire teetering structure on the verge of collapse, we find ourselves largely unable to reconceptualize "the economy" as being fundamentally intertwined with the whole of the Earth system, or to conceive of an economy not founded on ideals of perpetual growth. Furthermore, the radical free-market ideals that grew out of the postwar period envision that government too should join nature in getting out of the way as much as possible—setting some rules, perhaps, but then leaving the workings of the economy to "the market," by which theoretically is meant the aggregate decision-making of citizens. As a result, the economic thinking that has dominated most nations, and certainly the United States in recent decades, is unable to countenance the sort of centralized collective action that is needed to deal with a global catastrophe.

But to those inside the paradigm, including virtually all CEOs and high-level politicians, the limited view has generally appeared grand. Indeed, throughout much of the second half of the twentieth century economists, caught in their neat conception of circular flow, had a lot of trouble explaining why the US GDP was growing so fast. It was as if the wondrous machine they had built was performing far, far better than expected, and so each new uptick in GDP served as further proof that the machine was the best possible way to run an economy. The prominent economist Robert Solow, calculating in the 1950s, was so flummoxed by the US economy's rocket-launch trajectory that he resorted to ascribing more than five-sixths of the previous decades' economic growth to a murky category he called "technical change."[7]

It certainly is true that technological development can cause an increase in economic activity. But what Solow and other economists failed to factor in was readily available energy. Not until 2009 did a thorough analysis point out that the vast majority of twentieth-century economic growth in the United States and other countries could be ascribed directly to the availability of cheap energy, especially fossil fuels.[8] They were, it turned out, the magic elixir of economic growth, all the more potent because they, like all the most foundational of myths, were virtually invisible. "Its ready availability, in ever-increasing quantities, and mostly at relatively low and stable prices, meant that oil could be counted on *not to count*," Timothy Mitchell writes in his book *Carbon Democracy*, a detailed analysis of how the particular physical forms of fossil fuels have shaped politics worldwide.

It could be consumed as if there were no need to take account of the fact that its supply was not replenishable. In turn, not having to count the cost of humankind using up (largely within the space of two or three centuries) most of the earth's limited stores of fossil fuel made another kind of counting possible—new kinds of economic calculation. . . . The availability of abundant, low-cost energy allowed economists to abandon earlier concerns with the exhaustion of natural resources and represent material life instead as a system of monetary circulation—a circulation that could expand indefinitely without any problem of physical limits.[9]

In the United States, and in other former European colonies such as Canada and Australia, the turbocharging effect of fossil fuels was further complemented by the presence of vast quantities of other raw materials, including both cheap labor in the form of slaves (or their near equivalent in the form of Indigenous peoples or indentured immigrants) and abundant land. The latter was, to a large extent, readily rid of its previous human inhabitants, many of whom seemed weirdly uninclined to participate in the exciting prosperity gospel preached by their new neighbors. The land itself, tamed by mines and fences and roads and railroad tracks, constituted a storehouse of cheap resources and a safety valve for the pressures of burgeoning populations. In the case of the United States, historian William Appleman Williams called it "the escape hatch of the frontier," describing the bounty of available land as "a cast of mind as well as a stretch of open territory. This attitude has been undeniably exhilarating in a psychological and philosophical sense, and has led to beneficial consequences in a materialistic calculus."[10]

Land in its openness also provides an effective metaphorical way of understanding what is perhaps the most pernicious of economic myths, a sort of mathematical trick that truly does elevate our collective decision-making to the level of tragedy. It's called discounting, and it is at the root of nearly all decisions made about investing and public policy. At its core, discounting is a basic idea centering on versions of this seemingly simple question: If I've got a dollar for you, would you like it now, or a year from now? When psychologists or economists have posed this question to research subjects, the trend is consistent: *I want it now.*[11] It's a decision that makes both psychological and economic sense. I might, after all, be dead in a year, or otherwise unable to use the dollar; furthermore, I likely have ideas about how to use it now. Psychologically, we live much more intensively in the present than in the future, and so the dollar today has a hard reality, while the year-off dollar remains an abstraction. And economically it's highly rational to want the dollar now. I could invest it rather than spend it, and in that case it will be working for a year, earning interest or in other ways accruing value for me. If I use it wisely, I'll end up with more than a dollar next year. Furthermore, inflation dictates that a given amount of money generally declines in value over time. In a year, that same dollar is likely to have less purchasing power than it has now. It's an assumption that has been true throughout virtually all modern US history, and most of that of other developed nations too. The dollar now simply embodies more value than in the future, both psychologically and in real economic terms.

From these fundamental realizations about how both the mind and currencies work, economists have crafted an elaborate infrastructure of cost-benefit analysis intended to provide guidance on how much money it makes sense to spend now in order to achieve some kind of benefit in the future. That's discounting. As an economic tool it has done its job well through the decades and centuries of the expansion of Western economies; in fact, you might think of it as a sort of measurement tool for the American Dream. In an expanding economy it makes a lot of sense to value a future dollar less—even far less—than a current one, because it has generally been a safe bet that there are going to be more dollars around in the future. The Trump administration, for example, set a discount rate of 7 percent in calculating future benefits, meaning that preventing a dollar's worth of damage fifty years in the future is worth 3 cents of investment today, but no more.[12] Discounting has been the perfect mathematical expression of an ideal that has for many generations characterized the dominant American mythos, namely, that life for our children is going to be better, more comfortable, than it has been for us. Our present economic activity will see to that because it will create a bigger economy in the future, with more for all. "When ecology insists on the existence of limits, economic sciences find a way to invent a limitless future," writes Bruno Latour in *Down to Earth*.[13]

This assumption of growth is the same one that says that the GDP must grow, because that's what has almost always happened, and it has crept its way so thoroughly into policy thinking that it truly has an unseeable, mythic quality. Consider the *Stern Review,* an exhaustive study of the economics of climate change issued in 2006 by a team led by the prominent British economist Sir Nicholas Stern and subsequently published by Cambridge University Press. It caused a stir because Stern—no lefty environmentalist, but very much a member of the economics establishment—identified climate change as "the greatest market failure the world has ever seen."[14] The runaway trajectory of climate change, Stern's team wrote, is proof that the free market cannot police itself as neoliberal economists and conservative politicians like to imagine; rather, significant intervention by central governments is vital in addressing the problem. But even this critical report, widely regarded as a sober and groundbreaking analysis, was based on assumptions that the world economy would continue growing at least through 2050. "Even with climate change, the world will be richer in the future as a result of economic growth," it concluded.[15] The report then proceeded to use discounting—albeit at a very low level—to analyze just how devastating climate change is likely to be to the world's future GDP.

The logical outcome of discounting, one that is expressed in countless economic analyses of projects in infrastructure, health care, environmental protection, and numerous other realms of public life, is that if you look far ahead enough in time, you can calculate the future value of goods and resources as zero. If you assume that society will inevitably be richer in the future, then you can also assume that society's future members will find a way to muster the resources to deal with the problems we have caused for them. This is a very comforting way to look at things if you want to make money in the present, as most people do, and push the consequences off into the future, as is both psychologically and economically convenient. As the ethicist Chris Groves has written about the connection between societal risk-taking and climate change, "If we assume everyone will be better off in the future, then there is no incentive to act responsibly now."[16]

To grasp how seductive discounting is as a mathematical expression of how our society thinks about the future, I find utility in thinking of it in terms of visual metaphor rather than pure numbers. Imagine that the massive collective entity that we refer to as The Economy is a train hurling down the tracks—the same tracks we saw in chapter 1, with the *now* of climate breakdown seemingly ever receding into the distance, making it almost impossible to sense as a present danger. Think of discounting as a way of measuring the tracks, or more specifically the all-important width of the railroad ties. In the present, the ties have obvious size and heft—we can readily feel how important they are as the train hurtles its enormous weight over what seems a solid rail bed. But look forward: the ties appear to diminish in size until they vanish entirely. Artists use techniques of perspective to replicate this illusion in their paintings, tricking the eye into seeing three dimensions where there are really only two. Economists do the same thing, positing that dollars diminish in value as their appearance ticks further out into the future. We can't see them, we can't feel their importance, and once they're far enough away we assign them no value. That's true both for dollars and for the resources that the dollars represent. Your grandchild's right to a stable climate, to a Colorado River that's a reliable water source, to forests that aren't tinderboxes? It's worth paying 2.5 cents now for a dollar's worth of those values, according to how the Trump administration did its cost-benefit analysis. But 3.5 cents? That's a deal for a sucker, one that impinges on our right to make as much money in the here and now as possible.

In the American Southwest, one vivid illustration of the form discounting takes on the land is a literal set of railroad tracks, an eighty-mile route connecting a now defunct power plant set above the shoreline of Lake Powell with a once productive coal mine on Black Mesa to the east. Coal was mined on the mesa for more than fifty years and shipped in railcars to the Navajo Generating Station, whose electricity was needed to pump Colorado River water across hundreds of miles of desert to the booming Phoenix area. And there was enough coal to feed another power plant too, this one in southern Nevada. To transport coal all that way, Peabody Coal constructed a system that mixed pulverized coal with water, then piped it more than 270 miles across northern Arizona. The system used a billion gallons a year of scarce groundwater, and over decades dried up numerous springs that had once provided water for wildlife and for local Navajo and Hopi people. But it was profitable. The owners of the power plant paid lease fees of only $600,000 a year to the Navajo tribe, while

Peabody Coal paid the Hopi and Navajo tribes royalties at a rate one-fifth of those paid on federally managed public lands.[17]

The budget negotiations were a bad deal for the tribes, but at least how bad they were can be calculated. The drying up of springs that cannot be replenished, on the other hand, was not held to be of sufficient value to show up in budgeting decisions at all. Yet because much of the groundwater in the Southwest is "fossil water"—deposited during wetter periods in the past—those springs are likely to remain lost to many future generations of local people, who will have gained zero benefit but plenty of cost from those long-ago decisions.

The effect of discounting calculations, then, is that not only is a given date fifty years from now viewed as being of much less value than today. It's that the width of the few railroad ties that we rely on today and in a small set of years to come becomes more important than the width of all the ties we can see stretching far away from us, meaning that the economic value of the present few years—say, those until the next presidential election—outweighs *all future dates added together*. The value of resources for generations currently alive thus becomes more important than the value of resources *for all future generations*. So far we've been able to get away with this, mostly, because we simply assume from past experience that the railroad ties are going to be there when we need them. Surely someone will see to that. Someone always has. It's been such a safe assumption for sufficient decades, for sufficient generations, that it takes a mighty effort to acknowledge that the ties are in fact getting shorter and that soon they're not going to be sufficient to support us. The train really is going to come off the rails.

It's easy to defend discounting as an expression of a bedrock truth about how humans perceive their surroundings in space and time. When we're looking down a set of railroad tracks, they really do appear to meet off in the distance, even as we know they continue on over the horizon. And it really does appear to be a fundamental human quality to value money—or resources of any sort—in the present over those in the future; that assessment is not limited to contemporary hypercapitalist cultures. But it is worth remembering that even if these perceptions are firmly rooted in how our senses work, there is no inevitability in basing *our actions* on a particular predominant way of seeing. It is worth remembering that artists in other parts of the world do not to this day use the same rules of perspective that Western artists do, and that even Western artists did not adopt the rules of visual perspective that we today take for granted until the Renaissance. Before that, they would have painted a railroad in a very different way—likely with all the different cars the same size even if that's not the way they appear to our eyes in real life. Realism in Western painting is a myth in the sense that it is a set of rules that we have come to perceive as immutable truth, just as the ancient Greeks took for granted that tragic stories had to follow their predictable trajectories.

There is nothing inevitable about seeing the world in that way, and there is nothing inevitable about calculating that goods and resources must be worth less in the future than they are now. Other worldviews grant far more power to the future than prevailing western worldviews do. The best known is the seven generations idea inherent in various Native American traditions: when considering the impact of a contemporary decision, it is vital to consider the impact as far as seven generations into the future—without a discount rate. In this worldview a generation that is not to come for many decades carries the same weight, has the same value, deserves to have its fundamental needs met in the same way as the one that has to make the decision. In this worldview the idea of discounting the future, and the generations that will live there, is insanity.

It is an idea that has even arisen within Western economics itself. In 1928 a young Englishman and intellectual prodigy named Frank Ramsey published a paper titled "A Mathematical Theory of Saving" that explored, in dense, equation-rich text, how a people should plan for the future. He argued that individuals and societies should devote far more of their current resources to future use than was generally the case. The practice of discounting "later enjoyments in comparison with earlier ones," he wrote, "is ethically indefensible and arises merely from the weakness of the imagination."[18] Unfortunately, his promising ideas dropped with a thud, partly because Ramsey died very young in 1930 and partly because economists and politicians saw little reason to debate such questions of long-term planning when they had to deal with the more pressing matters of the Great Depression, World War II, and the postwar buildup. It's interesting to speculate what Ramsey might have done had he lived long enough to see the advance of the Keeling Curve and to understand climate change as the ultimate example of discounting the future.[19] It's also interesting to note that, according to a recent report from the International Monetary Fund, individuals and institutions whose investments fuel the global economy do not pay sufficient attention to future climate change risks even when those risks are viewed from a solely financial perspective.[20]

Even if they don't get much airtime, the existence of worldviews and of economics calculations that don't engage in discounting the future is a stark reminder that the operating system according to which mainstream society runs consists not of absolute truths but of constructs that we all have a hand in accepting or rejecting. But it is fair to ask whether all this truly amounts to a tragedy—with *tragedy*, remember, implying some foreknowledge of the flaw that is going to carry the plot into ruin. Have we really known enough, for long enough, about how climate breakdown has grown out of our economic decision-making that it is justifiable to call what's happening a tragedy?

In a word, yes. But to establish that I have to take some time at this point to disaggregate that pesky word *we*. *We* is a dangerous word to use in a consideration of a hugely complex problem of perception and policy. *We* stands very much at the heart of my analysis, but not in the sense of a word with a single meaning. It carries multiple meanings. To say that *we* have known about the impending breakdown of the world's climate for at least three decades is true only in a qualified sense when I use *we* to refer to the human species in the aggregate.[21] It is true that for at least three decades the elite institutions that most cultures have for generations relied on to make high-level decisions—governments, industry, academia, military planners—have known perfectly well that climate change is a real phenomenon that poses extraordinary risks to the continuance of healthy ecological processes and human communities. Representatives of those institutions have had access to all the information they needed to underscore that knowledge. As a species, *we* have known. But we cannot readily perceive being part of a species any more than we can truly perceive from our backyards that we are part of an entire globe, and so this is not at all the same as saying that *we all* have known. The disjunct between these two meanings of the word is at the core of the tragic narrative about climate breakdown and what to do about it.

One of the core theoretical arguments that has regularly been trotted out in support of neoliberal ideology—and it too is a myth—is that free markets work because everyone has equal access to equal information. Everyone decides what to produce and what to consume based on a collective understanding of what will work economically. All citizens have the right, and the power, to make their own economic decisions, to curate their own stories. Of course outcomes will be very different—some will grow rich, some will grow (or more likely remain) poor—but the process is fundamentally fair because everyone has access to the same information. In this ideal world it makes sense for government, with its heavy hand whose actions are likely based on political expediency, to get out of the way. In this ideal world we are all equal, in potential if not in economic outcome. In this ideal world *we* really would mean *all of us*.

But government has never gotten out of the way, and it was never meant to; nor were the powerful interests that have molded the neoliberal story all along; nor were the baked-in inequalities in information that are integral to a complex world. The powerful institutions that have shaped and governed the neoliberal order have proclaimed equality in responsibility and in potential even as they have used money and influence to assure that some are in fact very much more equal than others. They have known very well that their core narrative is a fiction even as they have promulgated it as gospel. It is this self-interested deception, rather than any fundamental flaw in human nature, that has made decisions on behalf of all of humanity. It is the powerful actors who have benefited from the free-market fiction who have, in promoting economic models based on discounting the future, all but assured that the future we will actually have resembles that in the models—a world greatly reduced in value for human and other life because its fundamental systems have been so thoroughly plundered for their present value. In so diligently projecting that future resources are worth zero when you look at them from the all-consuming perspective of the present, they have made it far more likely that they *will* in fact be worth zero because present-day economic activity is so rapidly draining value from the future.

It is here, in other words, that *we* breaks down into *us* and *them*; and what from some perspectives look like simple flaws in how humans think—how *we* think—becomes tragedy because it has all along been a plan carried out by only a small subset of people whose actions are affecting everyone. In the case of climate breakdown there is no question of ignorance, for those who have willfully steered us at full speed into the breakdown have known. They have known even as they have used ignorance as a weapon aimed at everyone else. We might argue, then, as I archly did in chapter 1, that this detailed lode of information and understanding has done more harm than good by undermining the core tenet of faith required in democratic societies, namely, that the widespread possession of quality information ultimately leads to quality decision-making. But what is equally relevant to the discussion of tragedy is that humanity's detailed understanding of climate breakdown has in fact had large and measurable effects—less on public policy than, tragically, on profits, as governments and corporate interests alike have used their knowledge of climate change for private rather than public good.

They have done so most notoriously in the case of fossil fuel corporations and allied interests that have funded elaborate denial campaigns, which might be described as smogscreens intended to distract the public from a full understanding of the problem while the deniers are busy carrying out their real work—of making money by developing tar sands, drilling in a rapidly melting Arctic, building more long-distance pipelines, and in myriad other ways continuing to squeeze profits from what should long ago have been a dying industry. It's been estimated that the profits of the global fossil fuel industry since 1990 have amounted to trillions of dollars—more than $2.4 trillion in 2019 dollars for the six largest Western-owned companies alone, according to an analysis by Taxpayers for Common Sense.[22] That number is far from the sum of overall profits for fossil fuel companies worldwide, let alone their allies in other industries, but it does function as a signpost showing the magnitude of the benefit that has been realized, by a relative few, in compensation for the enormous costs of climate breakdown present and future. It also tells us something about an incentive for crafting a particular public narrative: since 2000, five major Western oil companies have invested in the public and political support needed to sustain this flow of profits by spending well over a billion dollars a decade on advertising, much of it aimed at making the argument that untrammeled fossil fuel extraction and protection of the environment can be perfectly congruent.[23]

But denial, and fossil fuel profits, are only the most obvious ways in which information about the breakdown of the global climate—perhaps the ultimate example of a public good that ought to be owned or controlled by no one—has been used for private gain. The development of private information about climate change has itself become a growth industry. Recent years have seen a massive growth in "climate services" providers that charge steep fees to assist companies in understanding how they might adapt to and profit from climate change. Climate change consulting has become a multibillion-dollar industry whose chief characteristic is that the data consultants generate are private, and aimed at providing a basis for private profits—a far cry from the publicly accessible data furnished by the IPCC, the National Weather Service, and various other public-facing entities.[24] As the geographer Svenja Keele wrote in a 2019 analysis, "An emphasis on climate services shifts the incentives for climate science away from the public interest towards more commercial imperatives and in most cases a logic of profitmaking. Climate services seem likely to undermine the knowledge required for societies to adequately respond to the scale, speed and severity of climate change."[25] Climate services data, Keele points out, are far more likely to be used for adapting to climate change than to mitigating the problem on a global scale, and their proprietary nature suggests that this adaptation remains a private matter.

This form of information, in other words, carries embedded within it the implication that climate breakdown is not a matter for global collective action but rather for individualistic response—a neoliberal non-solution to a problem that has been greatly exacerbated by neoliberal thinking in the first place. It is a continuation of the long and dishonorable tradition of what Naomi Klein in her book *The Shock Doctrine* terms "disaster capitalism," or "disaster apartheid in which survival is determined by who can afford to pay for escape." Perhaps, she suggests, "part of the reason why so many of our elites, both political and corporate, are so sanguine about climate change is that they are confident they will be able to buy their way out of the worst of it."[26]

But if humanity has sold out its future for a few decades' worth of profits for fossil fuel companies and the many subsidiary industries and institutions that share their values, that's a cheap sort of tragedy expressing a shallow character flaw, namely, simple greed. Is that sufficient as an explanation for the question posed by Greta Thunberg with which I began this book? I propose that it is not, aware that what I am really writing about here may be no more than my own addiction to story, my hungry conviction that such a catastrophic breakdown in a species' inability to plan for the sustainability of its own future must have a deeper meaning, a richer and more resonant narrative linking the plot points.

So let us go further. Let us view climate breakdown—the ultimate ecological hyperobject—as the by-product of another hyperobject, namely, power. Like capital, power in its centralized form is a human creation that tends to take on a life of its own, that functions largely to perpetuate itself, that greedily seeks to exert ever more control over those who created it. Centralized power inherently needs to express and replicate itself. Compelling public narratives are among the most potent tools it uses to achieve that end. But even though it can be formulated into a grand narrative, there is in the end no story to it at all, no meaning that can be expressed in words. Hannah Arendt came to this conclusion in a 1963 letter in which she attempted to explain her controversial conception of the "banality of evil": "It is indeed my opinion now that evil is never 'radical,' that it is only extreme, and that it possesses neither depth nor any demonic dimension. It can overgrow and lay waste the whole world precisely because it spreads like a fungus on the surface. It is 'thought-defying,' as I said, because thought tries to reach some depth, to go to the roots, and the moment it concerns itself with evil, it is frustrated because there is nothing. That is its 'banality.' Only the good has depth and can be radical."[27]

The ultimate tragic way of understanding climate breakdown, then, is that it results from the human tendency to exercise power, and from the understanding that power exists not as a disembodied ideal but rather only in its expression. Centralized power has to exert itself regularly because otherwise it ceases to exist; like a virus, it cannot long survive outside its host in some free-floating form. It expresses itself most purely when its expression defies thought, has no rational goal. Dostoevsky knew this, creating in *Crime and Punishment* the character of Raskolnikov, who commits murder simply because he can, because to engage in a deeply consequential but senseless act is the ultimate means of expressing freedom. "Just as evil proclaims itself as a way of being free in the social world (the literary exemplar is the figure of Raskolnikov)," Clive Hamilton has written, "so the wanton neglect of the natural world is a sublime expression of our freedom. . . . To achieve this form of freedom one must decide to revel in one's autonomy without owning it, without taking responsibility for it."[28] Climate breakdown is the ultimate end result of the millennia-long Western project of embodying in action the supposed duality of humans and nature, an insistent demonstration of human freedom from limits—one that only increases in eloquence when it is carried out in the knowledge that there is in fact a huge price to be paid, that the limits are closing in fast.

It is no surprise that this battle cry of freedom should have reached its highest pitch in the United States. Inaction on climate change is on the surface a demonstration of power over nature—*we create our own reality*—but when it inevitably crashes into nature's limits it becomes a demonstration of another sort of power, one that in the American pantheon of values has come to hold higher value than life itself: the power of freedom. Just as millions of Americans reject plans for shared nationwide health care, even though its lack may kill them, and just as many of those same citizens reject the wearing of masks that might save their loved ones from dying in a pandemic, our society is behaving as though demonstrating our freedom by transgressing as we please against the stability of the global climate is a God-given right, not only one that we *can* exercise but one that we *must* exercise if we are to be free.

In his 2019 book *The End of the Myth,* the historian Greg Grandin explores why so many Americans appear to identify freedom most strongly with its crassest expressions. Trumpism, he wrote,

> encourages a petulant hedonism that forbids nothing and restrains nothing—the right to own guns, of course, but also to 'roll coal,' for example, as the rejiggering of truck engines to burn extraordinary amounts of diesel is called. The plume of black smoke emitted by these trucks is, according to such hobbyists, a 'brazen show of American freedom'—and, since 2016, a show of support for Donald Trump. Pulling out of the Paris Climate Accord did little to boost corporate profits, as many have pointed out, but it has everything to do with signaling that the United States will not succumb to limits.[29]

Seen in this light, climate breakdown becomes the epitome of the American ideal that we can do as we wish, or that we *have* to do as we wish, regardless of the outcome. And even though it's a tendency that one hopes reached its apogee with Trump, it's one that predates his emergence on the national political scene and has survived his reluctant departure from the nation's highest office. Its legacy certainly lives on in the disproportionate impact of US greenhouse gas emissions compared to those of other nations[30] and in the country's numerous expressions of what the political scientist Cara Daggett has termed "petro-masculinity."[31] Consider it an irony of history that the country most inclined to see a new morning always over the horizon should be the same one that has done the most to put future mornings into question for itself along with the rest of the world.

But if the truest expression of power is in a literal sense dumb, as Arendt implied—meaning that in resisting analysis it really no longer needs words at all—then that suggests that if anything can save us, it has to be the human gift for story, and in particular story as an expression of imagination. What unifies tragic narratives of climate breakdown, whatever their particulars, is that they are all failures of imagination. They are written by economists who see no point in envisioning life a generation into the future, by politicians who won't look beyond the next election, by capitalists obsessing over quarterly profit margins, by pundits unable to conceive of a world in which "the economy" does not have primacy over human life, by smart academics who use hopelessness as a measure of their braininess, by citizens whose brittle definition of freedom relies on shallow cultural signifiers like driving gas-guzzlers or refusing to wear a mask during a pandemic. Above all, as Joseph Meeker wrote, they embody the timeworn assumption that "man is essentially superior to animal, vegetable, and mineral nature and is destined to exercise mastery over all natural processes, including those of his own body. . . . Tragic art, together with the humanistic and theological ideologies upon which it rests, describes a world in which the processes of nature are relatively unimportant and always subservient to the spirit of man."[32]

But in any circumstance of addiction it is hardly a wise idea to slavishly follow the cues of the most addicted, those whose lack of ability to imagine a way out fills them with a perverse know-nothing pride. Thus I propose that we can best respond to the tragedy inherent in these narratives of climate breakdown by resisting the notion that a tragic narrative predetermining the outcome is an appropriate way to imagine the problem at all.

To do that, we need to focus on the second word as much as on the first. The word *narrative* implies a conscious creative effort constructed around the sense of an ending, an authorial intent whose outcome is clear before any audience is engaged. Narrative is the shared understanding that Oedipus must marry his mother and kill his father. It is the fatalistic belief that because fossil fuels are convenient and profitable we must use them all. It is the gospel of growth at all costs. It is the facile ideological math that says that if a problem would require a collective solution, it must not be acknowledged to be a problem. It is the conviction that the worse things get for the many, the closer we come to a longed-for culmination for the chosen few. It is the lazy assumption that we must follow the well-worn path of increasing tension until we are saved by a hero engineer or entrepreneur—or until everything comes crashing down around us. It is the satisfying schadenfreude the well-educated feel when they tell themselves that the future's dire arrival will show the unenlightened how wrong they have been. These all constitute narratives because narratives are honed and targeted stories that gain meaning from their endings. They are all narratives that express what the literature scholar Marek Oziewicz has termed "the ecocidal unconscious," which he describes as the process by which "even narratives envisioned to defend the biosphere tend to get twisted into narratives that reinforce the anthropocentric myopia which is destroying the biosphere in the first place."[33]

In the case of climate breakdown, these are narratives of power because they are all formulated by particular groups in government, business, religion, academia, or the media that have their own agendas to follow, their own selection processes for deciding what to include and what to exclude, and their own reasons for excluding other possible story lines. Narrative, the historian William Cronon has written, "cannot avoid a covert exercise of power: it inevitably sanctions some voices while silencing others."[34] A narrative is a spotlit pathway through an endlessly tangled thicket of events, an inevitably delimited understanding of the world. Labeling a particular explanatory scheme a narrative—however compelling or inevitable it is made to appear—implies that we are ignoring other possible modes of understanding. And the top-down societal narratives that prevail around climate breakdown invite—no, *require*—passivity of their audience; in accepting their framing of what's going on, we abandon our own faculties of observation and analysis. As the sociologist Nina Eliasoph wrote about American civic apathy at the close of the twentieth century, "Power works in part by robbing the powerless of the inclination or ability to develop their own interpretations of political issues."[35] Sure, Americans are free to make up their own minds about whether to "believe in" climate breakdown or not, or how seriously to take it, but given the power of the media and of big business, and our socialized hunger to fit ourselves into the beliefs of one group or other, the only choice many citizens see themselves as having is which ready-to-wear narrative to adopt. When the topic at hand is one as difficult as climate breakdown, the temptation to take the easy way out, to pretend that choice is out of reach, becomes almost irresistible. It is with a sense of relief that members of the public consign themselves to belief in one accepted climate narrative or another, on the broad spectrum that stretches from outright denial to hopeless fatalism. It doesn't even matter how dismal the content of the narrative is; embracing it is easier than accepting the complexities of choice and agency, responsibility and possibility.

No surprise, then, that the adolescent willfulness of much recent conservative activism—from rejecting the Paris Climate Accords to "rolling coal"—is more indicative of resignation to a particular ideology than of true freedom. In the classic formulation of the psychoanalyst Erich Fromm, such actions represent a "freedom from" rather than a "freedom to"—a resistance to limits and to regulation that is almost entirely uninformed by forward-looking thinking about what freedom might best be used *for*.[36] Like a seventeen-year-old's decision to drive fast without a seat belt simply because he newly has the ability to do so, they ultimately bespeak not the ability to shape one's future but its opposite—namely, the perverse and irresistible mandate to court harm because the possibility is there, a slavish need to follow a predetermined narrative. The direness of the consequence *is* the point, for it becomes the measure of our "freedom from." And so even as it embodies the narrative of tragedy, climate breakdown becomes an eerily precise embodiment of the United States' exceptional nationhood.

THE DEGROWTH ALTERNATIVE

The term "economy" stems from a Greek word referring to "management of the household." Climate breakdown is a clear sign that the household whose welfare ultimately matters most to future human prosperity and even survival is not the sort of local or national economy that has historically been the concern of policymakers; rather, it is the global one. Every additional degree of global warming, or even tenth of a degree, is further evidence that we have not learned how to live in our household in a way that would keep us within our means.

One marker of that inability is the slant of economic news in the mainstream media, which continues to focus undue attention on such misleading indicators as the GDP or the supposed success of the stock market. Yet an increasing number of economists are asking searching questions about how to maintain global economic health. Some are advocates for "decoupling," by which they mean efforts to ensure that economic growth no longer proceeds in lockstep with increases in greenhouse gas emissions, as it has for several centuries (figure 3.2). Through rapid development of renewable energy and other technological innovations, they argue, it ought to be possible for nations and communities to experience economic growth while simultaneously decreasing emissions.

It's possible to find evidence for this claim. For example, emissions of carbon dioxide (but not methane) in the United States and some Western European nations have been trending downward throughout much of this century so far, while GDPs have continued to grow.[37] Analysts ascribe this disjunction largely to reductions in those countries' use of coal. If it's possible for economic growth to continue while emissions decrease, isn't it possible that further development of renewable energy and other efficiency technologies could allow us to have the growth we're used to, without cooking the planet?

It appears not. What economists call the decoupling of growth and fossil-fuel emissions is almost as misleading an indicator as the GDP itself. There are two principal reasons for that. Industrialized nations have achieved their declines in emissions in part by outsourcing more-polluting or energy-intensive components of their economies to other nations where labor is cheaper or environmental regulations less stringent.[38] When Apple manufactures many of its computer components in China, the emissions associated with that work are counted as part of China's budget, not that of the United States or of other nations where the computers are bought and used. Yet governments don't take this globalization impact into account when they are developing national emissions-reductions targets.

Second, decoupling cannot happen quickly enough to allow for meeting any credible emissions targets. The economic anthropologist Jason Hickel and the economist Giorgos Kallis have argued that the most optimistic possible assumptions about how quickly societies can decarbonize their energy production and manufacturing fall far short of cutting emissions enough to meet the targets that the IPCC says are necessary to avoid catastrophic warming.[39] "A growth-obsessed economy powered by clean energy will still tip us into ecological disaster," Hickel writes.[40]

Hickel and Kallis have become two of the intellectual leaders of the "degrowth" movement, which seeks to shift the focus of the world economy from narrowly focused financial indicators such as GDP to the promulgation of policies that actually improve peoples' lives. They and others say it's actually fortunate that decoupling can't solve the climate crisis for us, arguing that the growth-oriented global economy has already done enormous damage in the form of economic inequality, resource colonialism, and pollution of all sorts. To reduce the scale of emissions while retaining the same economic systems and incentives that have placed us where we are now would be to continue the same behaviors that have wrought so much harm in ecosystems and communities worldwide.[41]

Degrowth, instead, seeks to measure economic vitality by how much it minimizes environmental harm and benefits the quality of life for all people. It would do away with measures like the GDP that focus unduly on financial metrics, and replace them with such tools as the Index of Sustainable Economic Welfare, developed by Herman Daly and John Cobb in the late 1980s, and the Genuine Progress Indicator, developed by Clifford Cobb in the 1990s at the think tank Redefining Progress to more fully quantify the welfare of nations. Leaders of nations from New Zealand to Costa Rica to China have announced steps in this direction.[42] Degrowth efforts emphasize such community-oriented ideas as promoting a sharing economy, worker cooperatives, and local production of food and energy.[43]

Degrowth analyses posit that the United States and other rich countries could afford significant drops in overall wealth while improving well-being, as long as income and such public goods as health care and education were much more equitably distributed than they are now.[44] That's a sign that big reductions in greenhouse gas emissions could not only complement a better quality of life for most. They could help produce a new narrative of national purpose, one focused on happiness, meaning, and healthy communities rather than on numbers, on thriving rather than on growth, as Kate Raworth told the journalist Eric Holthaus.[45] The goal, Hickel writes, is "a different kind of economy altogether—an economy that doesn't need growth in the first place. . . . We can create an economy that is organised around human flourishing instead of around endless capital accumulation; in other words, a post-capitalist economy. An economy that's fairer, more just, and more caring."[46]

5

COMEDY AND COMPLEXITY

Humans may not have begun their history in a state of primordial innocence, but they do appear to have begun it with a self-conscious aversion to being told what to do. . . . The real puzzle is not when chiefs, or even kings and queens, first appeared, but rather when it was no longer possible simply to laugh them out of court.

—David Graeber and David Wengrow[1]

In 1987 the novelist Joyce Carol Oates published a sardonic piece titled "Against Nature" in an anthology of essays about nature. It was an unlikely contribution to a volume full of rhapsodies on the virtues of staying put, of admiring the commonplace, of granting a full measure of grace to the present moment, most of them written by authors who might safely be described as residents of the somewhat obscure literary suburb condescendingly termed "nature writing."[2] That neighborhood was too bland for her, Oates wrote. Nature, she claimed, "has no sense of humor." It lacks both "a moral purpose" and "a symbolic subtext." Perhaps worst of all for storytellers, it "lacks a satiric dimension, registers no irony."[3] It simply couldn't take center stage in the sorts of dramas that readers want, which need to be about characters who can appreciate humor, experience conflict, exercise or shred morality, or play with satire and irony so that readers can with a flush of superior pleasure recognize those abilities in themselves. Nature, as part of the stage set represented by climate, was simply too static to be of dramatic interest.

Bemused or irritated as her co-essayists might have been, deliberately snarky as she might have been, Oates was on to something. For much of modern society, nature as we have generally defined it has not been dramatic, or ironic. It has not been moral; worse, it has not been immoral. What it has been is boring. Not to everyone: there are always those pockets of people with specialized interests in birds and prairie wildflowers and the squirming residents of tide pools, but as far as mainstream society is concerned, nature has been variously a storehouse of resources, an entertaining place for outings from more civilized places, an endless source of beautiful images, and sometimes a threatening force that wallops us with storms and wildfires. In the modern technological society in which Oates was writing before climate breakdown became a major concern, nature—or climate—did not have the dramatic heft to play more than a bit part.

Above all, nature could not be tragic. Whatever else might be said about heroes or villains of the sort with which Oates and so many other storytellers have populated their works, they are never boring—at least not if the story is going to sell. Nature, on the other hand, we have viewed as fundamentally lacking in the qualities of tragedy, especially in its lack of allegiance to grand ideals. We have viewed nature as simply *being*, flourishing but dully void of self-examination. No wonder we have perceived living within nature's limits as dull also, preferring the flash and buzz of moving beyond them, whatever the cost. As the ecological economist Nicholas Georgescu-Roegen suggested in the 1970s, "It's as if the human race had chosen to lead a brief but exciting life, leaving to less ambitious species a long but monotonous existence."[4]

As a safe climate breaks down, though, nature is becoming less boring every day, in a way that is pretty much synonymous with *terrifying*. Unabated, the path of climate breakdown is indeed chilling, and tragic. But it is exactly in our recognition that this narrative is explicitly tragic that we might be able to find a way out.

To explore that, I need to return to Joseph Meeker, whom I introduced in the previous chapter. Meeker is interesting, largely because he is not only a literary scholar but also a student of nature; he worked as a field biologist and park ranger for many years. As a result, he was uniquely well situated to read stories about how people face problems—or fail to—from the perspective of how nature deals with problems. His realization, as expressed in *The Comedy of Survival*, was that nature is not only not tragic but the *opposite* of tragic: it's comic. And it is through comedy that we might envision a way out from under the thumb of the varied narratives of tragedy that keep us from addressing climate breakdown.[5]

To Meeker, tragedy can be fueled by any one of a number of Big Ideas such as "mastery" and "greatness."[6] Comedy, on the other hand, has a single goal: survival. It "grows from the biological circumstances of life"; it "demonstrates that man is durable even though he may be weak, stupid, and undignified"; it "is a celebration, a ritual renewal of biological welfare as it persists in spite of any reasons there may be for feeling metaphysical despair"; it "is concerned with muddling through, not with progress or perfection."[7] Meeker finds a striking parallel between human traditions of comic storytelling, often expressed in slapstick or picaresque progressions of ridiculousness, and the colorful and unpredictable workings of evolution. "Productive and stable ecosystems are those which minimize destructive aggression, encourage maximum diversity, and seek to establish equilibrium among their participants—which is essentially what happens in literary comedy," he writes. "Biological evolution itself shows all the flexibility of comic drama, and little of the monolithic passion peculiar to tragedy."[8] Even though it centers on the practice of perpetual change, evolution is the perfect counterpart for what in the realm of human relations with place we have come to call *sustainability*, or the long-standing contract between generations that those currently in charge will not constrain the future opportunities of those who are not yet able to exercise their own political power.

Tragedy, we might say, centers on the practice of saying *yes* to a narrowly circumscribed set of ideas and by doing so saying *no* to a far larger set; comedy, like evolution, functions by not saying *no* to anything. As the scientist-turned-filmmaker Randy Olson points out in his entertaining memoir-cum-communication guide, *Don't Be Such a Scientist*, improvisational comedy has only one rule: *Never say no*.[9] It is a showstopper for a player in an improv sketch to say *Don't be ridiculous—that could never happen*. And that same rule is the very one evolution follows. It is telling that when the first dried and stuffed specimens of platypuses began showing up in Europe in the nineteenth century, scientists thought that some mischievous taxidermists in Australia were having a joke at their expense.[10] What else could explain the existence of an aquatic mammal with webbed feet and a duck's bill whose males sported a venomous spur on their hind legs, whose females laid eggs? (It would take until 2020 before biologists realized that platypuses and some other Australian marsupials are even stranger than was known back then, for they also glow blue under ultraviolet light.[11] Why? Good question.) These specimens did not fit the accepted narrative of taxonomy. But of course, it is exactly by trying endless solutions to the problems of survival, without a preordained narrative plan, that evolution succeeds in ensuring the persistence of life.

By suggesting that our response to climate breakdown should be comic, I am not implying that we should merely joke about it—though that actually is a good start. The climate communication expert Max Boykoff and the theater and environmental studies scholar Beth Osnes suggest in a recent paper that while "science is often privileged as the dominant way by which climate change is articulated, comedic approaches can influence how meanings course through the veins of our social body, shaping our coping and survival practices in contemporary life. . . . The dynamism and non-linearity of comedy provides potential sites of powerful resistance within adversity."[12] Humor can be subversive, as the veterans of numerous campaigns to oust or reform autocratic governments can attest; it can also help promote better public understanding, even of such challenging topics as climate change.[13] And by centering such attitudes as irony, playfulness, and irreverence, says the environmental humanities scholar Nicole Seymour, it can be "particularly suited to address such horrors" as climate change.[14]

But the comic tradition encompasses much more than what we readily identify as humor. Meeker holds comedy to include such literary traditions as the picaresque, finding in John Yossarian of Joseph Heller's *Catch-22* an unlikely antihero who, by valuing above all his own survival in an absurd wartime situation, is able to transcend the hierarchical power structures that constantly threaten to send him to his death. Though the reading experience may be comic, the character's experience is far from funny in any lighthearted sense.

What we can take from this interpretation of what the comic is, and from nature's tradition of evolution, is the understanding that it is not foreordained that we make ourselves subject to a dominant narrative of human progress, of economics, of what is portrayed by a narrow set of powerful narrators as being "politically feasible." Comedy has always been deeply democratic. It has always fed off the practice of breaking down boundaries—between humans and nature, men and women, the privileged and the oppressed. Comedy holds that any sort of people can find agency, if not by grabbing political power, then by mocking it. The long tradition of comedy proposes that we make ourselves subject to narratives of power, or of what's "feasible," only if we choose to do so—and if we do, that is the real tragedy. Comedy traffics in endlessly varied possibilities, in crafting stories of meaning that, as Meeker writes of ecology, "challenges mankind to vigorous complexity, not passive simplicity."[15] If evolution can remake a mammal by having it lay eggs and putting a bill on it, can't humans remake a destructive economy that they themselves created?

Even the ancient Greeks, as they were developing their tradition of dramatic tragedy, recognized its inherent potential for tyranny, whether political or emotional. They incorporated into performances of tragedies a humorous interlude—typically a satyr play that allowed performers and audience to experience release through bawdiness, wordplay, and sexual suggestiveness.[16] They also softened the rigidity of tragic narratives themselves by having a chorus accompany the main action. The chorus was basically a citizens' commentary that served to explicate the characters and the plot; it is believed to have grown out of still older performative rituals of the sort that persist among many land-based, agricultural peoples. It represented the *demos*, or the people, and thereby served as a bridge between the audience and the plot. It was an exercise in democracy. By providing interpretation and perspective that shed new light on the core action, the chorus delivered a real-time, street-level analysis of the tragic, and its observations were regarded as critical to the experience of catharsis, the ultimate payoff for sitting through the sorrows of a tragic play.[17]

We too need catharsis. We—and here I mindfully do mean *all of us*—need to conceive of the climate breakdown story as one that incorporates a much more varied and multifarious stream of voices than has generally been the case. Unlike narrative, drama does not have to be mapped out in advance; that's why we use the word *dramatic* to convey the sense of not knowing where a story is headed. Unlike narrative, drama is continuously flexible, adaptive; its path can be altered by what the actors decide to do. Unlike tragic narrative, the outcome of drama is not predetermined by suprahuman forces or preexisting conditions. As the theologian-biologist Celia Deane-Drummond has written,

> Narratives, when told in a certain way as grand narratives, create an aura of determinism, where what is anticipated seems an almost inevitable trajectory of the story as told so far. . . . For narrative it is possible to remain detached and distant from the events being described. Drama is much more engaging in that at minimum it invites the audience to get involved by imagining that they are one of the players in the drama. Where identification is complete, a drama allows the audience to be caught up as participants in the drama itself, so that a deep sense of individual agency is evoked.[18]

When it comes to climate change, Deane-Drummond posits, switching from a narrative to a dramatic frame could have enormous consequences for the sense of efficacy that citizens feel: "A focus on the dramatic would lend itself to a more active sense of responsibility. It would imply that the inevitable march of history in epic narrative is not all there is to be said, for we are participants as much as observers."[19] Perhaps, by envisioning climate breakdown as a drama involving all of us rather than as a tragedy preordained by the interests of a few, we might develop the agency—the active participation—that's needed to turn its course in a more hopeful direction.

Because of the violent history of our species, there are ample precursors for a society's survival hinging on a comedic or adaptive approach to life, even if most people in those societies would not view their situations as comic at all. It is not surprising that many come from communities or societies whose survival has been threatened by those with greater economic or military power.[20] As the essayist Mary Annaïse Heglar has written, climate change is far from the first existential threat that Black people have faced in the United States.[21] And the same is true for Indigenous peoples, who have for centuries been criticizing the Western assumptions about individualism and economic growth that have done the most to get us into this mess.[22] "The idea that disaster will come is not new," writes the feminist scholar Donna Haraway in her discussion of how Navajo people on Black Mesa have responded to violence and colonization; "disaster, indeed genocide and devastated home places, has already come, decades and centuries ago, and it has not stopped."[23] If it is incumbent on those of us with greater privilege in today's society to do what we can to ameliorate those ongoing disasters, it would also be prudent for us to learn something of the possibility of survival through tough conditions from those with long experience of suffering.

The philosopher Jonathan Lear writes of finding an example of this attitude in the Crow people of the western Great Plains as they faced the onslaught of colonization and expropriation in the late nineteenth century. Losing a primary source of sustenance—enormous free-ranging herds of bison—and their traditional lands, losing even the culturally meaningful practice of engaging in small-scale warfare with the Sioux, the Crow were at risk of cultural extinction. Skilled warriors, they could have chosen the path of noble tragedy by fighting until the end. But they did not. Under the guidance of an exceptional leader, Plenty Coups, who himself had been guided by an evocative dream or vision featuring a chickadee, they decided to make themselves pliable. They decided to give up many of the practices of diet, warfare, measurement of status, and other markers that had made life in their culture meaningful, even though this sacrifice in many ways meant giving up their identity. "The tribe's problem was not just that they did not know what the future had in store," Lear writes, "they lacked the concepts with which to experience it."[24] The Crow were facing "the breakdown of a culture's sense of possibility itself."[25] As they made decisions that would convert them from nomadic hunters to a sedentary culture that adopted some of the new settlers' traditions of farming and ranching, the Crow were moving into a future for which they lacked a framework of ethical practice and understanding. Their "comedic" approach was to do so with the faith that a new framework would emerge over time, even if its outlines could not be perceived or even imagined when the crisis began. One might call it an act of societal improvisation.

According to Lear's analysis, the obscured but hopeful view of the future that he reads in the Crow tribe's response to cultural upheaval is an ideal analog for the situation all humanity finds itself in now. It centers on the same point I have been making, namely, that at a time of crisis, undue adherence to preset and readily understandable narratives is a sure path to ruin. It's possible to move forward, Lear writes, only through "radical hope," through "a commitment to the idea that the goodness of the world transcends one's limited and vulnerable attempts to understand it. There is no implication that one can glimpse what lies beyond the horizons of one's historically situated understanding. . . . Radical hope anticipates a good for which those who have the hope as yet lack the appropriate concepts with which to understand it."[26]

The urgency of moving forward while changing course, without knowing the precise outlines of what we are moving into, is now so pressing that there ought to be a word for it. Fortunately, at least two have been suggested. The philosopher Kathleen Dean Moore has invented one, as she suggests in her book, *Great Tide Rising*: "*presilience*, literally, forward-jumping, the courage to take a great stumbling leap into a world unlike any we have ever seen, knowing that we will not be returning to the old ways."[27] It is a process, she argues, that cannot be scripted in advance but must be created along the way, using all the tools and voices of truly democratic decision-making. "No one has written the Save the World Symphony," she writes. "We're making it up as we go along. This is not classical music. This is jazz, with all its risk and glory."[28] Another candidate is "hopepunk," coined by the writer Alexandra Rowland to describe an attitude of caring and brash independence from oppressive norms: "Hopepunk isn't ever about submission or acceptance. It's about standing up and fighting for what you believe in."[29]

To move mindfully into a murky future, to embrace its unknownness, to practice presilience or hopepunk, is to reject the tragic narrative even if we know that much of what is to come will be terrible. It is to find the hope that lies "between disaster and catastrophe," as the British scholar and climate change activist Rupert Read has phrased it.[30] We will certainly face climate disasters; we already have. But the hope that can avert catastrophe lies in the realization that tragic narratives, for all their very real power, are brittle. Like railroad tracks, they cannot readily change course when the need arises. Like towers of concrete and steel, they seem inevitable and permanent—until suddenly they are not. On September 11, 2001, the world watched as the myth that the United States could engage in endless entanglements overseas without violent consequences at home shattered in a shocking and unexpected way. Through the exercise of radical hope, the woven strands constituting the myth that the climate breakdown story is heading toward some tragic or glorious or inevitable end can snap too, victims of their own inflexibility. And then we might be freed, as the communication scholar Julia Corbett has written: "Right now, we are at an impasse between the old stories of fossil fuel culture and new ones yet to be written. Moving through this mythic story is akin to a rite of passage: separating from the old story and reckoning with the past, progressing through a transition that is both liminal and dangerous, and finally reincorporating fully into new ecologies and economies."[31]

Democratic comedy, radical hope, presilience, hopepunk—these frameworks for action center the persistence of life itself just as evolution does. They are snarky. They are slippery. They can be sexy. They are the seed sprouting through concrete, the hermit crab living in a shell of discarded plastic. Like a platypus, they embrace contradiction and complexity. Requiring thinking outside any box, they are representative of an adaptive attitude toward life that is ever slipping the narrow yoke of definition. They might best be seen as an expression of life protecting itself from forces that, unchecked, would lead to its demise or degradation.

In the case of climate breakdown, acting in the spirit of radical hope is an example of what the Australian writer James Button calls "politics—not of the left or right side, but of the body."[32] To act as if there can be a promising future even if we can't see it, even if we fear it may not be there at all, is to embody a deep truth: that we have been animal for far longer than we have been human. The life force within us, valuing as it does its own persistence, is the result of billions of years of evolution, of nature doing its thing. Maybe Joyce Carol Oates was right on a small scale when she claimed that nature "registers no irony." But on the largest of scales nature has certainly been an ironist, creating through those eons of unnarrated change a species that finally developed the ability to craft its own narratives, only to see many of them become parasites that drain more and more potential from the future in pursuit of their present-day coherence. Much of today's difference in attitudes toward climate breakdown, and toward nature itself, comes down to whether a person or group of people views this odd twist of Earth's history as irony—or as the purpose of the whole exercise.

I hew toward the former view because irony lends itself to malleability in a way that an overarching purpose never will. We do love telling and living through stories. But here's the rub: to prioritize the persistence of life even in the face of overwhelming odds and seductive tragic narratives always falls short of being fully explainable by a cohesive story because it is fundamentally a preverbal act. It cannot entirely be put into words because it is the product of four billion years' worth of evolution rather than of a few hundred millennia of human culture. To remember this is to remember also that what is at stake in the climate crisis is not just particular human lives or communities or ways of organizing ourselves; it is also polar bears, and coral reefs, and ponderosa pine forests, and the duck-billed platypus, and monsoon rains, and the carving of canyons by rivers, and myriad other expressions of how life has worked through evolution to fill our planet's surface—a gazillion relatives, in other words. It is to feel an allegiance to where we came from that has perhaps never been better expressed, at least in the traditions of Western literature, than in Henry David Thoreau's barely coherent expression of what he felt on descending the wild flanks of Mount Katahdin in Maine: "Talk of mysteries!—Think of our life in nature,—daily to be shown matter, to come in contact with it,—rocks, trees, wind on our cheeks! the *solid* earth! the *actual* world! the *common sense! Contact! Contact! Who* are we? *Where* are we?"[33]

The sense within us that feels a deep affinity for life rebels at the prospects we increasingly see before us, at the threats to so many of evolution's most vibrant manifestations. As Bill McKibben phrased it in an interview with the journalist Emily Atkins, "We've lobotomized that interesting quarter of God's brain that is the coral reef."[34] There is no path before us that does not entail uncountable losses. Our apprehension of these losses may come to us largely through images and words and numbers, yet it resonates within us at a nonverbal level that cannot be reconciled with human-centered stories designed to rationally explain away the losses, or to grasp at increasingly unlikely means of averting them. That we are all in denial to some degree certainly constitutes an effort at avoiding both the sting of responsibility and the pain of helpless looking-on. But it is also a reminder that we are all responding to a stimulus that transcends our human capacity for story. It grounds us as animals whose ancestors surely knew the feelings of caring and loss long before they developed words for them—and who knew that when danger comes, it is often brisk action that is needed, before the articulation of the tale.

This might seem an expression of inadequacy, especially coming from a writer with a deep love of the study and practice of story. But it is also a realization that points to a way forward, for humans have always grappled with the contradictions that form part of the endless negotiation between place and culture. Healthy societies have always had to acknowledge that nature controls them, sort of, even as they control nature, sort of. They have always known that they are composed of beings who are both animal and beyond-animal. They have had to contend with the knowledge that to live, they have to take from Earth, sometimes bloodily. To accomplish these matters of relationship, they have used the time-tested technologies of art and music and poetry and dance, all of them tied to ordinary language in some way but all moving beyond it to some other space in which the working out of relationships results in the creation of meaning.

This is a practice, or set of practices, that is well known in Indigenous communities that retain a tie to place, and that has also come to be increasingly understood by those who seek to heal places anew. "Words can go only so far as tools in the work of transmuting fact into truth," writes the ecological restoration expert Bill Jordan III, who has argued that those seeking to restore threatened ecosystems need to take a cue in building such relationships from the ritual traditions of land-based peoples. "For that, another link in the chain may be necessary. To the act that converts data into facts must be added the ritual through which facts are *act*-ually transmitted into truth, beauty, and meaning."[35] It is the *acting* that is important, he argues, in both senses of the word: humans working out complex relationships with land and place *act out* roles that provide purpose to the *actions* they have to take to live. The physical effect on land and plants and animals is important, of course. But so is the less tangible effect on the human actors, who find that their seemingly externally focused work has acted on *them*: it produces new understandings about what it means to be simultaneously a cultural and a physical animal, one with a hunger for meaning as well as for calories. Acting, Jordan writes, opens up a liminal space that, by embracing contradiction, acknowledges the tangled knot of responsibilities and story lines within which cultures have always found themselves.

What we need, then, is the act of restoration, of renewed relationship, but scaled up to encompass the entire Earth. We have to acknowledge that we cannot restore the planet to anything like the health and diversity our distant ancestors encountered; our losses are already large, and they will only accumulate. But we can experience and embody the power, the creation of meaning, that comes of acting, of getting off the couch and away from the screen, of going to the neighborhood meeting, of raising a voice or a shovel. Action transforms. It transforms not least the actors, changing their perspective on what they see before them, opening up new vistas of possibility, of story. It can transform audiences, too, showing rather than telling them that there are positive steps to be taken. Writ large, the aggregation of many smaller actions can transform entire societies. Knowledge can shape action, but it's equally true that action shapes knowledge. I'm pretty sure Greta Thunberg knows that. So do many others who through action are finding their own powerful voices—and slowly, collectively, giving form to the idea that it is in working against climate breakdown that humanity can create a new and more hopeful story, even if the ending is unknown.

In "The Carrier-Bag Theory of Fiction," Ursula Le Guin writes that a story starring a spear or a spear-wielder is a tale of triumph and tragedy, of imposing one's will through a projection of power—of overcoming engineering problems, we might say. Such stories quite literally come to a point. Stories conceived according to the idea of gathering, on the other hand, aim neither toward "resolution nor stasis but continuing process." Their purpose is continuation: passing knowledge on to the next generation to ensure it has the same tools we received from those who went before. This "life story," as she calls it, comes in the form of "myths of creation and transformation, trickster stories, folktales, jokes, novels."[36] What such tales have in common is that they build the future not by writing over the past, as triumphal narratives of progress do, but by embracing it. They find meaning in chronos in ways that tragic narratives cannot.

What we need, then, is not to discard narrative but to reclaim it from the dominance of the relative handful of narrators whose dead-end story lines have taken us to the brink of catastrophe. We need to expand the world's narrative, the human story, from a few inflexible strands of meaning into a woven tapestry of far more threads. We need to broaden it so that it does not incessantly look only forward, as the dominant Western narratives of progress and of apocalyptic endings have tended to do. We need to broaden it so that it melds the difficult truths of grief and mourning and exultation with the more familiar discourses of science and policy. We need to broaden it so that it encompasses the full complexity of relationships between humans and the more-than-human, the world of plants and animals and weather and climate cycles that is increasingly reclaiming its place as an actor. And we need to broaden it so that it contains within it the voices of far more people of far more cultures than have so far been represented in mainstream discourse and decision-making.

Indeed, we can reclaim the idea of a vibrant future only by reclaiming a broad sense of that word *we*. In their 2020 book *The Upswing*, Robert Putnam and Shaylyn Romney Garrett examine multiple trends that illustrate how the United States has evolved since the mid-twentieth century from a society focused on the collaborative to one focused on the individual. They find in the expansive Progressive movement of the late nineteenth and early twentieth centuries a good model for how Americans can transition from a current overly individualistic understanding to a more collective sensibility—one not wedded to consistency. "The movement was so diverse as to be barely coherent," they write, "and was home to contradictory impulses."[37] But the result of having such a diverse movement, they propose, was a broad societal shift that saw widely accepted narratives move away from "I"-centered thinking and toward "we"-centered perceptions of society. Those narratives generally retained prejudiced thinking about nonwhite Americans. But they did succeed in ameliorating numerous ills, from inequality in education and wealth to the oppression of women.

The stories that will grow out of climate breakdown also need that sort of breadth more than they require coherence. As storytellers and as actors, we need the unanticipatable sparks that come of rubbing against one another in myriad new ways. "We require each other in unexpected collaborations and combinations, in hot compost piles," writes Donna Haraway of the power of juxtaposition, of adding on to rather than subtracting from.[38] Like improvisational comics, we should be spending more time saying *yes* than saying *no*—a response that has become more reflexive than ever in the echo chambers of social media. Like evolution, we need above all to embrace multiplicity, to leave ourselves open to the possibility that some unexpected and productive responses to climate breakdown may come in mash-up forms as weird as the seeming amalgam of life forms that we call a platypus.

Such stories can come in all forms and in all sorts of venues. The writer Kendra Pierre-Louis saw one in *Black Panther*, finding in the film's portrayal of the fictional Wakanda a past- and future-focused linkage of pastoralism and high technology, one that sustainably melded the rural and the urban even as it reimagined possible relations between races. "We need different stories, ones that help us envision a present in which humans live in concert with our environment," she writes. "One in which we eat, play, move, and live in ways that are not just lighter on the Earth but also nurturing to us as humans, with at least some of the trappings that many of us have come to expect of modern life."[39] What if, she asks, "the story we tell about climate change is . . . an opportunity? One for humans to repair our relationship with the Earth and reenvision our societies in ways that are not just in keeping with our ecosystems but also make our lives better?"[40]

We do have powerful tools for this evolution, or revolution. Bruno Latour writes in *Down to Earth* that the only way out of humanity's climate predicament is through a conjoining of two means of self-identity that have almost always been seen as polar opposites. What humanity has to accomplish, he argues, is "carrying out two complementary movements that the ordeal of modernization has made contradictory: *attaching oneself to a particular patch of soil* on the one hand, *having access to the global world* on the other. Up to now, it is true, such an operation has been considered impossible: between the two, it is said, one has to choose. It is this apparent contradiction that current history may be bringing to an end."[41] Thanks to climate change, the local and the global are at last becoming one.

For much of modern history, as Latour indicates, this would never have been considered possible: if to be local meant caring deeply about place, it was also to be provincial, narrow, self-interested; if to be global implied having a certain broad-mindedness, it was also to be disconnected, elite, a snob looking disdainfully down at flyover country. But we now have both the technology and the urgency to overcome this old divide, to become "eco-cosmopolitans," as the environmental humanities scholar Ursula Heise has phrased it.[42] We can readily learn from the Wakandas of the world and from their opposites, the countless places that have been grievously harmed by the same processes that are changing the climate. We can compare notes, trade ideas, share strategies, create the new *Black Panthers* that point a way forward. We can benefit from entire bagsful of stories that grow out of specific places but point out to the wider world, that help build a better future by bringing forward the lessons of the past. We can fit a lot into our gathering sacks, at least the kind of utilitarian carrier-bag Le Guin has in mind.

In its commodiousness, ready to take advantage of whatever is ripe today, the bag itself can serve as a reminder that in looking for any single, particular narrative of how to resist climate breakdown, we might well be overlooking other possibilities. To spend too much time looking for coherence can be to hamstring ourselves. To expend undue effort on worrying about the effectiveness of particular climate change solutions, to argue endlessly about which approach is the right one, can come to resemble those medieval theological arguments about angels dancing on the head of a pin. Even then it was far more meaningful, and efficacious, to treat God as God in a more straightforward way, to accept that there were rather simple reasons to behave in particular ways that had nothing to do with final outcomes or with the details of doctrine.

Earlier I suggested that the most meaningful possible analogy for climate change is God. I can't say I'm not being snarky in doing so, but the suggestion contains this kernel of truth, or of encouragement: for thousands of years, uncounted adherents of various religious traditions have undergone all sorts of extreme sacrifice in the here and now on the supposition that it was the right thing to do given the existence of some sort of deity or deities, or some sort of broad-based belief system transcending individual lives. The value of these actions could never be fully assessed by their effect in the moment, nor could it be known for certain whether the sacrifice would bring some sort of reward after death. But the sacrifice was still worth making, the martyrdom worth taking. For believers it was, and is, worth acting as if God exists, whatever the cost of that embodied belief might be. And for all of us now, it is worth acting as if climate breakdown is upon us, even if we cannot know up front whether our actions in dealing with it will bring about any particular desired result. We need to act meaningfully even if we have no way now of knowing whether our actions will be successful, or whether there will be any future meaning in what we do.

Perhaps this is nature's biggest irony, the ultimate joke: if we manage to escape the climate trap we have set for ourselves through our undue allegiance to dead-end narratives, we will thereby be crafting the biggest story of them all, the one documenting how we managed to open up a new future of possibility.

To address climate breakdown as if the future matters is the greatest task humans have ever faced. An honest assessment at this point has to conclude that it is uncertain whether we are up to the job. But because we are the product of billions of years of biological evolution and hundreds of thousands of years of cultural evolution, it is our obligation to both our ancestors and our descendants to try. Our task is to step out of the yoke of the narratives whose comfortable weight we have allowed to settle on us over centuries and to find regenerative stories that link us to something new—and to something much older. As the journalist Jonathan Mingle has written, "In industrialized societies, we don't tend to think of ourselves as ancestors—in the deep, many-generations sense—but that's what the moment demands."[43]

Dealing with climate breakdown is ultimately going to be a story either of unfathomable human decline or—if we are to be good ancestors—of a movement into some not-yet-imagined future featuring more just and sustainable relations among peoples and between humans and the more-than-human world. One way or the other, it is going to be the ultimate story of humanity. No wonder so many have been trying to write it. Yet no one can. No *one*, that is. What happens will consist of myriad stories rather than of a narrative plotted in advance. When the time comes, new myths reflecting and governing humans' relations with one another and with the rest of the world will no doubt arise. Much as we might want to, we cannot pre-write them. Our job is not to write an account of the future. It is to allow for the ongoing possibility that that story, or that web of stories, can be written along the way in an unending sequence that links past and future. "We still have the chance to *make the space* for hope," writes the activist Emily Johnston, "to act in such a way that hope might exist for others who come after us."[44]

Here is one small step in the great and ongoing project of making space for hope, and for future stories. For almost five decades, the three giant smokestacks of the Navajo Generating Station in far northern Arizona were the tallest human-built structures in the American West, punctuating a desert slickrock landscape just above the azure waters of Lake Powell on the Colorado River. At the end of 2020, they came tumbling down in a controlled explosion, one by one leaning sideways and then tipping at what seemed a remarkably languorous pace into a massive cloud of dust that dissipated only slowly. The coal-fired power plant whose emissions they had ejected into the atmosphere—among the biggest sources of carbon dioxide in the country—could no longer compete with cheaper natural gas or with renewable energy producers. And with the shuttering of the power plant, the Kayenta Mine on Black Mesa, which had provided the coal, closed too.

The decline of the coal economy is good news, though whether natural gas is really much better from a climate standpoint remains an open question. But as environmentalists celebrated, and as Navajo and Hopi officials worried over how to offset the loss of hundreds of well-paying fossil fuel jobs on Native lands marked by persistent high unemployment, I thought instead of a story I'd heard a while back. It was about a group of coal miners returning home from a day at the mine. As they did every day, they commuted in an old school bus belonging to Peabody Coal. As they did every day, they had to descend a steep road swooping down from the mesa top on the way to their homes in the nearby rez town of Kayenta.

On this particular day, the brakes failed. The driver had no way to stop the bus. It barreled down the final straightaway toward the intersection with the busy highway that would take them to Kayenta—if they survived that long. The road continued on a level across the highway, but the drivers doing sixty-five and seventy on Route 160 would hardly be expecting an old school bus to shoot through a stop sign at reckless speed. The risk of a catastrophic collision with a semi truck or a vanload of tourists or even a real school bus was high.

The danger was real, the tension thick. And so the Navajo men in the back of the bus did the only thing they could think to do: they cracked jokes. It was better than sitting rigid with fear. Even if they couldn't know the outcome. Especially because they couldn't know the outcome.

Well, here is their story, one that opened the door for countless others. Because I am telling it, you know how things turned out. They were all in it together, and they survived.

They got through. We might too.

FOUR TIPPING POINTS TOWARD NEW NARRATIVES

Despair can easily send us in the wrong direction, but so can hope. We already know so much about the terrors of climate breakdown; the very act of reaching for hope that things will not be so bad can invest the physics of what is happening with more mystery, more uncertainty, than is warranted. And, like so many of the other mental processes we reach for to survive, that can mentally channel our thinking into the same well-worn narrative grooves in which an awful problem can be overcome through heroism, doggedness, or even luck.

But still. I cannot end without hope, which after all is an active verb as well as a state of mind. Nor can I fully content myself with what inevitably remain somewhat abstract calls to embrace the multiplicity and spontaneity of the comic, to practice the arts, to immerse ourselves in nature, to act even as we do not know what the outcome of our acting will or can be. And so I want to consider several phenomena or trends that have the potential to swiftly shift the boundaries of what is perceived as possible—and with it the embedded foundations for stories that can help illuminate and explain our path.

The first is technological innovations, especially those related to energy production. This is not to say that we can engineer or grow our way out of climate breakdown. To shift our entire societal infrastructure to run on renewable sources of energy without dramatically downsizing the amount of energy we consume would require ruinous quantities of raw materials. To suck sufficient carbon dioxide out of the atmosphere would itself require extraordinary amounts of energy, or the conversion of developed or agricultural land back to forest and other ecological carbon sinks at a time when those ecosystems are struggling to survive. But what innovative technology can do is dramatically change the economic playing field by removing power from fossil fuel companies and their allies; this, of course, is why fossil fuel interests and their political servants have thrown up so many hurdles. Those companies have been able to thrive as they have only through enormous subsidies in the form of tax breaks for developing new places to mine and to drill, taxpayer-funded construction of roads and other infrastructure, and a failure to pay the costs of the harm they have caused in the form of pollution, political corruption, and climate change. As electric cars compete more effectively with gasoline-powered ones, as more electricity is generated without fossil fuels, as grotesque ocean pollution reveals the heavy cost of plastic, the economic power of the sector that has done the most harm diminishes.

The second is legal liability. Fossil fuel companies have garnered those enormous profits and caused enormous harm while disclaiming all responsibility for climate breakdown and numerous local pollution disasters. To date they have paid little for the privilege of degrading the future. But that may be changing. Around the world, including in the United States, legal cases seeking to hold the industry accountable for the climate breakdown they knew was coming are proceeding. Though some lawsuits have failed, others will follow. It's likely that the discovery process in some of these cases will force the industry to reveal what its own scientists and officials knew about climate change even as corporations were funding disinformation campaigns. If one or more of these cases document this information gap in black and white, the doors to enormous financial liability open wide. And at that point, the industry will quickly lose a great deal of political influence, just as happened with the tobacco industry in the United States once it was found liable for the health consequences of smoking. An industry facing enormous liability penalties and diminished profits will no longer be able to buy influence from elected officials as it has for so long. When that happens, the political playing field may shift quickly.

The third is climate breakdown itself, perhaps the most certain of these phenomena. It may seem an expression of what the futurist Bruce

Sterling once called "dark euphoria" to find hope in the inevitable physical devastation and societal disruption of climate breakdown.[45] But with these awful consequences, with humanity's increasing immersions in the clarifying experiences of kairos, will come rapid possibilities for new narrative understandings of humanity's situation and future. These possibilities include the potential for tribal, authoritarian, even fascistic political responses. But they can also, as the essayist Rebecca Solnit has explored in her book *A Paradise Built in Hell*, include a renewal of democratic and egalitarian approaches that could address some of the immediate consequences of climate breakdown while also promoting the development of more just and resilient societies and communities.[46] By striking at many of the pillars whose support for today's lifeways is already growing increasingly tenuous, climate breakdown as a physical phenomenon will inevitably change perceptions of what is politically necessary and societally possible. What will become apparent to rapidly growing numbers of people is that the struggle against climate breakdown is literally a fight for our lives, or at least for those of our children.

The fourth is the rapidly growing engagement of today's youth with the climate crisis. Greta Thunberg and millions of other young people are modeling what it can look like to embody resistance to the passive acceptance of climate breakdown—to live a new narrative unbound by the stale patterns of doing and understanding that older people have such a hard time breaking through. "It is counterproductive . . . to think of climate change as just an issue for Democrats or 'liberals,'" writes Sarah Jaquette Ray in her recent book, *A Field Guide to Climate Anxiety*, which is explicitly targeted at a young audience, "and far more important to begin mobilizing along generational lines."[47] It may be time, in other words, for young people to make the difficult decision that many have to make at some point in the lifetime of their family, namely, when the old folks are showing signs of dementia by squandering their savings or otherwise making bad decisions, it's time to take away the keys or claim power of attorney. A shift of political power toward young people has the profound potential to change the norms of how personal life narratives are shaped, and, by extension, the norms of what society deems culturally possible.

All these potential changes constitute what others have called a "social tipping element"—a perturbation in a human-centered system that can cause unanticipated change to take hold fast, analogous to the tipping points that the physical world may be facing.[48] They are reminders that climate breakdown is the opposite of an engineering problem, with its set flowcharts of cause and effect. It is, rather, an adaptive problem in which the entire playing field and the possibilities it entails shifts as new moves are made.

None of these evolutions alone is sufficient to avert uncounted tragedies. Even all of them together will not save us from more pain and heartbreak than we might think ourselves able to bear. But together they may be able to shift the trajectory of the future, to help us transition quickly from evolution to revolution. Together, they have the potential to establish the outlines of new narratives able to envision a future beyond climate breakdown—stories that, by pushing the frontiers of imagination, shape new possibilities of a sort we can't see from our current vantage point.

NOTES

INTRODUCTION

1. Anthony Leiserowitz, Connie Roser-Renouf, Jennifer Marlon, et al., "Global Warming's Six Americas: A Review and Recommendations for Climate Change Communication," *Current Opinion in Behavioral Sciences* 42 (2021): 97–103.

2. Anthony Leiserowitz et al., *International Public Opinion on Climate Change* (New Haven, CT, 2021: Yale Program on Climate Change Communication and Facebook Data for Good).

3. International Energy Agency, Sustainable Recovery Tracker (Paris, 2021), https://www.iea.org/reports/sustainable-recovery-tracker.

4. Among recent examples, see Katharine Hayhoe, *Saving Us: A Climate Scientist's Case for Hope and Healing in a Divided World* (New York: One Signal Publishers/Atria, 2021); Julia B. Corbett, *Communicating the Climate Crisis: New Directions for Facing What Lies Ahead* (Lanham, MD: Lexington Books, 2021); Rebecca Huntley, *How to Talk about Climate Change in a Way That Makes a Difference* (Sydney, Australia: Murdoch Books, 2020); and Elin Kelsey, *Hope Matters: Why Changing the Way We Think Is Critical to Solving the Environmental Crisis* (Vancouver, BC: Greystone Books, 2020). For a more scholarly overview, see Max Boykoff, *Creative (Climate) Communications: Productive Pathways for Science, Policy and Society* (Cambridge: Cambridge University Press, 2019).

5. David Wallace-Wells, *The Uninhabitable Earth: Life after Warming* (New York: Tim Duggan Books, 2019), 147.

6. Kerry Arsenault, "Storyteller at the Fire: An Interview with Jonathan Lethem," *Orion* 39, no. 1 (Spring 2020): 87.

7. Amitav Ghosh, *The Great Derangement: Climate Change and the Unthinkable* (Chicago: University of Chicago Press, 2016), 138.

8. Rob Nixon, *Slow Violence and the Environmentalism of the Poor* (Cambridge, MA: Harvard University Press, 2011), 14.

9. David Mackie and Jessica Murray, "Risky Business: The Climate and the Macroeconomy," J. P. Morgan Economic Research Report, January 14, 2020.

10. I am well aware of what risk scholars have termed the "White-male effect," which posits that some 30 percent of white males seem to have remarkably low sensitivities to risks that most others perceive as dangerous—and for the record, I don't think I'm among that 30 percent. For a discussion of this phenomenon, see Paul Slovic, "Trust, Emotion, Sex, Politics, and Science: Surveying the Risk-Assessment Battlefield," *Risk Analysis* 19, no. 4 (1999): 689–701.

11. I am far from the first to dissect the word "we" in this way with reference to climate change. Joel Wainwright and Geoff Mann's *Climate Leviathan: A Political Theory of Our Planetary Future* (London: Verso, 2018), for example, offers a Marxist-based critique of *we*, *us*, and *them* in exploring several possible future politics of a climate-changed world.

12. Timothy Garton Ash, *The Free World: America, Europe, and the Surprising Future of the West* (New York: Random House, 2004), 3.

13. According to a recent analysis, the United States has been responsible for 40 percent of "excess global CO_2 emissions" since 1850, with the EU nations responsible for another 29 percent. Jason Hickel, "Quantifying National Responsibility for Climate Breakdown: An Equality-Based Attribution Approach for Carbon Dioxide Emissions in Excess of the Planetary Boundary," *Lancet Planet Health* 4 (2020): e399–404.

14. For example, Naomi Oreskes and Erik Conway, *Merchants of Doubt: How a Handful of Scientists Obscured the Truth on Issues from Tobacco Smoke to Global Warming* (New York: Bloomsbury, 2011); Inside Climate News, "Exxon: The Road Not Taken," https://insideclimatenews.org/book/exxon-the-road-not-taken; Jane Mayer, *Dark Money: The Hidden History of the Billionaires behind the Rise of the Radical Right* (New York: Anchor, 2017); Dario Kenner, *Carbon Inequality: The Role of the Richest in Climate Change* (Abingdon-on-Thames: Routledge, 2019); Christopher Leonard, *Kochland: The Secret History of Koch Industries and Corporate Power in America* (New York: Simon & Schuster, 2019); John Cook, Geoffrey Supran, Stephan Lewandowsky, et al., *America Misled: How the Fossil Fuel Industry Deliberately Misled Americans About Climate Change* (Fairfax, VA: George Mason University Center for Climate Change Communication, 2019); Geoffrey Supran and Naomi Oreskes, "Rhetoric and Frame Analysis of ExxonMobil's Climate Change Communications," *One Earth* 4 (2021): 696–719.

15. For example, Paul Hawken, ed., *Drawdown: The Most Comprehensive Plan Ever Proposed to Reverse Global Warming* (New York: Penguin, 2017); Jonathan Safran Foer, *We Are the Weather: Saving the Planet Begins at Breakfast* (New York: Farrar, Straus and Giroux, 2019); Mary DeMocker, *The Parents' Guide to Climate Revolution* (San Francisco, CA: New World Library, 2018); Solomon Goldstein-Rose, *The 100% Solution: A Plan for Solving Climate Change* (Brooklyn, NY: Melville House, 2020).

16. Robert Jay Lifton, *The Climate Swerve* (New York: New Press, 2017), 148.

CHAPTER 1: PREDICTION

1. Jan Zwicky, "A Ship from Delos," in Robert Bringhurst and Jan Zwicky, *Learning to Die: Wisdom in the Age of Climate Crisis* (Regina, Saskatchewan: University of Regina Press, 2018), 69.

2. Hard-core climate change skeptics reading this may well use this metaphor in exactly the opposite way, claiming that the *now* we never reach is the dire future disaster "alarmists" like me make.

3. Peter Frumhoff, "Global Warming Fact: More Than Half of All Industrial CO_2 Pollution Has Been Emitted Since 1988," Union of Concerned Scientists, December 15, 2014, https://blog.ucsusa.org/peter-frumhoff/global-warming-fact-co2-emissions-since-1988-764.

4. Timothy M. Lenton, Johan Rockström, Owen Gaffney, et al., "Climate Tipping Points—Too Risky to Bet Against," *Nature* 575 (November 28, 2019): 592–595.

5. Stephen Schneider, *Global Warming: Are We Entering the Greenhouse Century?* (San Francisco, CA: Sierra Club Books, 1989).

6. Nathaniel Rich, *Losing Earth: A Recent History* (New York: MCD, 2019), 152.

7. Inside Climate News, "Exxon: The Road Not Taken," https://insideclimatenews.org/book/exxon-the-road-not-taken; Rich, *Losing Earth*, 171.

8. Philip Shabecoff, "Global Warming Has Begun, Expert Tells Senate," *New York Times*, June 24, 1988, A1, A14.

9. Wallace S. Broecker, "Climatic Change: Are We on the Brink of a Pronounced Global Warming?" *Science* 189, no. 4201 (Aug. 8, 1975): 460–463.

10. President's Science Advisory Committee, Environmental Pollution Panel, *Restoring the Quality of Our Environment* (Washington, DC: White House, 1965), 112–133.

11. Mike Hulme, *Why We Disagree about Climate Change: Understanding Controversy, Inaction and Opportunity* (Cambridge: Cambridge University Press, 2009), 53.

12. Hulme, *Why We Disagree about Climate Change*, 47–48.

13. Ayana Elizabeth Johnson and Katharine K. Wilkinson, *All We Can Save: Truth, Courage, and Solutions for the Climate Crisis* (New York: One World, 2020), xvii.

14. Marc Reisner, *Cadillac Desert: The American West and Its Disappearing Water* (New York: Penguin, 1986).

15. Reisner, *Cadillac Desert*, 255–257.

16. Chi Xu, Timothy A. Kohler, Timothy M. Lenton, et al., "Future of the Human Climate Niche," *PNAS* 117, no. 21 (May 26, 2020): 11350–11355.

17. Jay Griffiths, "Myths of Stability," *Orion* 34, no. 6 (November–December 2013): 13–14, 13.

18. Ralph F. Keeling, S. C. Piper, R. B. Bacastow, et al., "Atmospheric CO_2 Data," Scripps CO_2 Program, Scripps Institution of Oceanography, https://scrippsco2.ucsd.edu/data/atmospheric_co2/primary_mlo_co2_record.html.

19. The concentration of methane has increased too, owing to factors ranging from the digestive systems of cattle to leaks from natural gas extraction and shipping. So have concentrations of less common but in some cases more potent greenhouse gases such as nitrous oxide and chlorofluorocarbons.

20. Global Carbon Budget, Carbon Budget and Trends 2021, November 4, 2021, https://www.globalcarbonproject.org/carbonbudget/21/publications.htm.

21. Albert Szent-Györgyi, "The Brain, Morals and Politics," *Bulletin of the Atomic Scientists* 20, no. 5 (May 1964): 3.

22. Lijing Cheng, Kevin E. Trenberth, John Fasullo, et al., "Improved Estimates of Ocean Heat Content from 1960 to 2015," *Science Advances* 3, no. 3 (March 2017).

23. Climate scientists have pointed out that extreme winter cold spells in eastern North America and Europe are often paired with unusually warm weather in the Arctic, and are actually a signal of climate change. See Judah Cohen, Karl Pfeiffer, and Jennifer A. Francis, "Warm Arctic Episodes Linked with Increased Frequency of Extreme Winter Weather in the United States," *Nature Communications* 9 (2018), doi: 10.1038/s41467-018-02992-9.

24. Frances C. Moore, Nick Obradovich, Flavio Lehner, et al., "Rapidly Declining Remarkability of Temperature Anomalies May Obscure Public Perception of Climate Change," *PNAS* 116, no. 11 (March 2019): 4905–4910.

25. National Oceanic and Atmospheric Administration, "NOAA Delivers New U.S. Climate Normals," May 4, 2021, https://www.ncei.noaa.gov/news/noaa-delivers-new-us-climate-normals.

26. Daniel Pauly, "Anecdotes and the Shifting Baseline Syndrome of Fisheries," *Trends in Ecology and Evolution* 10 (1995): 430.

27. Andrew Nikiforuk, *Empire of the Beetle: How Human Folly and a Tiny Bug Are Killing North America's Great Forests* (Vancouver, BC: Greystone Books, 2011).

28. Peter D. Howe and Anthony Leiserowitz, "Who Remembers a Hot Summer or a Cold Winter? The Asymmetric Effect of Beliefs about Global Warming on Perceptions of Local Climate Conditions in the U.S.," *Global Environmental Change* 23, no. 6 (December 2013): 1488–1500. It is worth pointing out, though, that heat waves are the sensory signal that people seem to find the *most* indicative of global climate change effects. Survey respondents are more likely to associate their experience of hot weather with an acknowledgement that climate change affects them personally than other unusual weather signals such as changes in precipitation. See J. R. Marlon, Xinran Wang, Matto Mildenberger, et al, "Hot Dry Days Increase Perceived Experience with Global Warming," *Global Environmental Change* 68 (May 2021), art. 102247, https://doi.org/10.1016/j.gloenvcha.2021.102247.

29. Alison Flood, "Oxford Junior Dictionary's Replacement of 'Natural' Words with 21st-Century Terms Sparks Outcry," *Guardian*, January 13, 2015, https://www.theguardian.com/books/2015/jan/13/oxford-junior-dictionary-replacement-natural-words.

30. Griffiths, "Myths of Stability," 14.

31. Donald Trump to an audience of military veterans in 2018: "Just remember, what you're seeing and what you're reading is not what's happening." As George Orwell pointed out, instilling doubt in what our senses tell us is a primary tool of the autocrat.

32. Danielle Garrand, "President of Finland Denies Telling Trump the Country Rakes Its Forests to Prevent Fires," *CBS News,* November 18, 2018, https://www.cbsnews.com/news/finland-president-sauli-niinisto-denies-telling-donald-trump-the-country-rakes-forests.

33. Peter Friederici, ed., *Ecological Restoration of Southwestern Ponderosa Pine Forests* (Covelo, CA: Island Press, 2004).

34. Eleonora M. C. Demaria, Pieter Hazenberg, Russell L. Scott, et al., "Intensification of the North American Monsoon Rainfall as Observed from a Long-Term High-Density Gauge Network," *Geophysical Research Letters* 46, no. 12 (June 2019): 6839–6847; Salvatore Pascale, "Current and Future Variations of the Monsoons of the Americas in a Warming Climate," *Current Climate Change Reports* (2019), https://doi.org/10.1007/s40641-019-00135-w; Bin Wang, Michela Biasutti, Michael P. Byrne, et al., "Monsoons Climate Change Assessment," *Bulletin of the American Meteorological Society* 102, no. 1 (January 2021): E1–E19, doi: https://doi.org/10.1175/BAMS-D-19-0335.1.

35. P. C. D. Milly and K. A. Dunne, "Colorado River Flow Dwindles as Warming-Driven Loss of Reflective Snow Energizes Evaporation," *Science* 367, no. 6483 (March 2020): 1252–1255.

36. A. Park Williams, Edward R. Cook, Jason E. Smerdon, et al., "Large Contribution from Anthropogenic Warming to an Emerging North American Megadrought," *Science* 368, no. 6488 (April 2020): 314–318.

37. Hulme, *Why We Disagree about Climate Change*, 327–328.

38. Alastair McIntosh, *Riders on the Storm: The Climate Crisis and the Survival of Being* (Edinburgh: Birlinn, 2020), 126.

39. Ulrich Beck, *Risikogesellschaft: Auf dem Weg in Eine Andere Moderne* (Frankfurt: Suhrkamp, 1986), 96 (my translation).

40. Let's overlook, for now, that "gut reactions" occur primarily in the brain and not in the intestine. As the science storyteller Randy Olson points out in *Don't Be Such a Scientist: Talking Substance in an Age of Style* (Covelo, CA: Island Press, 2009), 20, the convenience of thinking of conceptions like *my gut told me* and *belly laughs* as connected to the body's literal viscera is too useful a shorthand not to use.

41. Jonathan Safran Foer, *We Are the Weather: Saving the Planet Begins at Breakfast* (New York: Farrar, Straus and Giroux, 2019), 34.

42. Zeke Hausfather, Henri D. Drake, Tristan Abbott, et al., "Evaluating the Performance of Past Climate Model Projections," *Geophysical Research Letters* 47 (2020): e2019GL085378, https://doi.org/10.1029/2019GL085378.

43. Bruce Hardy and Kathleen Hall Jamieson, "Overcoming Biases in Processing of Time Series Data about Climate," in *The Oxford Handbook of the Science of Science Communication*, ed. Kathleen Hall Jamieson, Dan M. Kahan, and Dietram A. Scheufele (New York: Oxford University Press, 2018).

44. Will Steffen, Johan Rockström, Katherine Richardson, et al., "Trajectories of the Earth System in the Anthropocene," *PNAS* 115, no. 33 (August 14, 2018): 8252–8259.

45. David Spratt and Ian Dunlop, *Existential Climate-Related Security Risk: A Scenario Approach* (Melbourne, Victoria: Breakthrough—National Centre for Climate Restoration, May 2019), 9.

46. Mike Hulme, "Four Meanings of Climate Change," in *Future Ethics: Climate Change and Apocalyptic Imagination*, ed. Stefan Skrimshire (London: Continuum, 2010), 44.

47. Climate Reality Report, "2030 or Bust: 5 Key Takeaways from the IPCC Report," October 18, 2018, https://www.climaterealityproject.org/blog/2030-or-bust-5-key-takeaways-ipcc-report.

48. These predictions have not been wrong. Climate meltdown did not abruptly start in 2016, but it is true that society's failure to change course by then had locked in so much future warming that terrible calamities are almost certain to result. And the IPCC contention that major changes have to be implemented by 2030 is true too. The problem is that the selection of a specific date or year as a precise deadline for either climate breakdown or policy change works against an understanding that the problem is one of incremental but accelerating change lacking easily identifiable inflection points.

49. Figure 1.1 adapted from Keeling et al., "Atmospheric CO_2 Data"; Jan C. Minx, Max Callaghan, William F. Lamb, et al., "Learning about Climate Change Solutions in the IPCC and Beyond," *Environmental Science and Policy* 7 (2017): 252–259.

50. Arendt, "Reflections: Truth and Politics," *The New Yorker*, February 25, 1967, 50, 54.

51. Howard Mansfield, *The Same Ax, Twice: Restoration and Renewal in a Throwaway Age* (Lebanon, NH: University Press of New England, 2000), 62.

52. Milly and Dunne, "Colorado River Flow Dwindles."

53. Daniel Sherrell, *Warmth: Coming of Age at the End of Our World* (New York: Penguin, 2021), 99.

54. David Biello, "Climate Negotiations Fail to Keep Pace with Science," *Scientific American*, December 7, 2011, https://www.scientificamerican.com/article/climate-negotiations-fail.

55. Masha Gessen, "To Be or Not to Be," *New York Review of Books*, February 8, 2018, https://www.nybooks.com/articles/2018/02/08/to-be-or-not-to-be.

56. Robin Wall Kimmerer, *Braiding Sweetgrass: Indigenous Wisdom, Scientific Knowledge, and the Teachings of Plants* (Minneapolis, MN: Milkweed Editions, 2013), 48.

57. Kathryn Newfont with Debbie Lee, "Introduction," in *The Land Speaks: New Voices at the Intersection of Oral and Environmental History*, ed. Debbie Lee and Kathryn Newfont (New York: Oxford University Press, 2017), 24.

58. Patrick Nunn, *The Edge of Memory: Ancient Stories, Oral Tradition and the Post-Glacial World* (London: Bloomsbury Sigma, 2018).

59. Lynne Kelly, *The Memory Code: The Secrets of Stonehenge, Easter Island and Other Ancient Monuments* (New York: Pegasus Books, 2017), 7–10.

60. Rory A. Walshe and Patrick D. Nunn, "Integration of Indigenous Knowledge and Disaster Risk Reduction: A Case Study from Baie Martelli, Pentecost Island, Vanuatu," *International Journal of Disaster Risk Science* 3, no. 4 (2012): 185–194.

61. Nunn, *The Edge of Memory*, 26–27.

62. Margaret Hiza Redsteer, Klara B. Kelley, Harris Francis, et al., "Increasing Vulnerability of the Navajo People to Drought and Climate Change in the Southwestern United States," in *Indigenous Knowledge for Climate Change Assessment and Adaptation*, ed. Douglas Nakashima, Jennifer T. Rubis, and Igor Krupnik (Cambridge: Cambridge University Press, 2018).

63. Peter Friederici, "Private Memories of Public Precipitation," in *The Land Speaks*, ed. Lee and Newfont.

64. Kimmerer, *Braiding Sweetgrass*, 346.

CHAPTER 2: METAPHOR

1. Jamais Cascio, "The Apocalypse: Not the End of the World," *Bulletin of the Atomic Scientists* 75, no. 6 (2019): 269.

2. Peter Goin and Peter Friederici, *A New Form of Beauty: Glen Canyon Beyond Climate Change* (Tucson: University of Arizona Press, 2016).

3. John Wesley Powell, *The Exploration of the Colorado River and Its Canyons* (Mineola, NY: Dover, 1961), 247.

4. The most comprehensive recent account of what happened to them is in Don Lago, *The Powell Expedition: New Discoveries about John Wesley Powell's 1869 River Journey* (Reno: University of Nevada Press, 2018).

5. Jay Griffiths, *Wild: An Elemental Journey* (London: Penguin, 2006), 37.

6. In grammatical terms, a comparison utilizing the word *like* is actually a simile, while a metaphor states that the more unknown object or idea being described equates to the more familiar one, as in "metaphor is a rope." But they share the same linguistic potential and shortcomings.

7. Luntz Research Companies, *The Environment: A Cleaner, Safer, Healthier America* (2002), https://www.sourcewatch.org/images/4/45/LuntzResearch.Memo.pdf, 142.

8. Anthony Leiserowitz, Geoff Feinberg, Seth Rosenthal, et al., *What's in a Name? Global Warming vs. Climate Change*, Yale University and George Mason University (New Haven, CT: Yale Project on Climate Change Communication, 2014).

9. Many years later, Luntz testified before a Senate committee on the seriousness of the climate crisis, saying, "I was wrong." Kate Yoder, "Frank Luntz, the GOP's Message Master, Calls for Climate Action," *Grist*, July 25, 2019, https://grist.org/article/the-gops-most-famous-messaging-strategist-calls-for-climate-action.

10. George Lakoff and Mark Johnson, *Metaphors We Live By* (Chicago: University of Chicago Press), 53.

11. Sadly, this was not entirely true as I was wrapping up this book: in July 2021 a rafter was killed and others badly injured when a flash flood that could not have been

predicted roared through a rafters' camp in Grand Canyon. Increased flash flood severity has been projected as a significant climate change impact in the Southwest. Gabriela Miranda, "Grand Canyon Flash Flood Leaves 1 Woman Dead, Others Injured," *USA Today*, July 17, 2021, https://www.usatoday.com/story/news/nation/2021/07/17/grand-canyon-flood-1-dead-others-injured-after-flash-flood/8001885002.

12. Xiang Chen, "The Greenhouse Metaphor and the Greenhouse Effect," in *Philosophy and Cognitive Science, SAPERE 2*, ed. L. Magnani and L. Li (Berlin: Springer, 2012).

13. Cynthia Taylor and Bryan M. Dewsbury, "On the Problem and Promise of Metaphor Use in Science and Science Communication," *Journal of Microbiology & Biology Education* 19, no. 1 (2018): 1–5.

14. Lakoff and Johnson, *Metaphors We Live By*, 12.

15. Judah Cohen, Karl Pfeiffer, and Jennifer A. Francis, "Warm Arctic Episodes Linked with Increased Frequency of Extreme Winter Weather in the United States," *Nature Communications* 9 (2018), doi: 10.1038/s41467-018-02992-9.

16. Mitchell Thomashow, *Bringing the Biosphere Home: Learning to Perceive Global Environmental Change* (Cambridge, MA: MIT Press, 2002), 26.

17. Bill McKibben, "When Words Fail," *Orion* 27, no. 4 (July/August 2008): 18–19.

18. Bill McKibben, "A World at War," *New Republic* 247, no. 9 (September 2016), 24.

19. Stephen J. Flusberg, Teenie Matlock, and Paul H. Thibodeau, "Metaphors for the War (or Race) against Climate Change," *Environmental Communication* 11, no. 6 (2017): 769–783.

20. The philosophers Kyle Fruh and Marcus Hedahl point out, though, that because climate change does present an existential threat to low-lying island nations, it can be construed as a form of unjust war perpetrated by wealthy nations on their poorer cousins: "Climate Change Is Unjust War," *Academia Letters* April 2021, art. 510, 1–3.

21. Lakoff and Johnson, *Metaphors We Live By*, 140.

22. John Pollack, *Shortcut: How Analogies Reveal Connections, Spark Innovation, and Sell Our Greatest Ideas* (New York: Avery, 2015), 151–153.

23. In *Metaphors We Live By*, Lakoff and Johnson also explore (156–157) how President Jimmy Carter chose to use a war metaphor in addressing the 1970s energy crisis, a linguistic choice that through its associations defined a particular set of policy options and—without debate—excluded others. Carter is said to have been encouraged to do so by an adviser who suggested that framing the issue this way would salve the nation's "crisis of spirit." Philip Hammond, *Climate Change and Post-Political Communication* (Abingdon-on-Thames: Routledge, 2018), 31.

24. See, for example, Supran and Oreskes, "Rhetoric and Frame Analysis of ExxonMobil's Climate Change Communications."

25. Timothy Morton, *Hyperobjects: Philosophy and Ecology after the End of the World* (Minneapolis: University of Minnesota Press, 2013), 20.

26. Amitav Ghosh, *The Great Derangement: Climate Change and the Unthinkable* (Chicago: University of Chicago Press, 2016), 115.

27. Rob Nixon, "Slang Is Changing at a Glacial Pace," Quartz, April 2, 2018, https://qz.com/1242923/slang-is-changing-at-a-glacial-pace-and-shows-how-blind-we-are-to-the-horrors-of-climate-change.

28. BBC, "UK Weather: What Are the Effects of a Heatwave?," July 2, 2018, https://www.bbc.com/news/uk-44680164.

29. George Monbiot, "Climate Change? Try Catastrophic Climate Breakdown," *Guardian,* September 27, 2013, https://www.theguardian.com/environment/georgemonbiot/2013/sep/27/ipcc-climate-change-report-global-warming.

30. Sophie Zeldin-O'Neill, "'It's a Crisis, Not a Change': The Six Guardian Language Changes on Climate Matters," *Guardian,* October 16, 2019, https://www.theguardian.com/environment/2019/oct/16/guardian-language-changes-climate-environment.

31. Rupert Read and Wolfgang Knorr, "This Is Not an Emergency . . . It's Much More Serious Than That," *Emerge,* February 21, 2022, https://www.whatisemerging.com/opinions/climate-this-is-not-an-emergency-it-s-much-more-serious-than-that.

32. Christoph Mauch, "Slow Hope: Rethinking Ecologies of Crisis and Fear," *RCC Perspectives: Transformations in Environment and Society* 1 (2019): 17.

33. Bruno Latour, *Down to Earth: Politics in the New Climatic Regime,* trans. Catherine Porter (Cambridge: Polity, 2018), 44.

34. John Schwartz and Richard Fausset, "North Carolina, Warned of Rising Seas, Chose to Favor Development," *New York Times,* September 12, 2018, https://www.nytimes.com/2018/09/12/us/north-carolina-coast-hurricane.html.

35. Tristram Korten, "In Florida, Officials Ban Term 'Climate Change,'" *Miami Herald,* March 8, 2015, https://www.miamiherald.com/news/state/florida/article12983720.html.

36. National Task Force on Rule of Law & Democracy, *Proposals for Reform,* vol. 2 (New York: New York University Brennan Center for Justice, 2019); Christopher Flavelle, "Conservative States Seek Billions to Brace for Disaster. (Just Don't Call It Climate Change)," *New York Times,* January 20, 2020, https://www.nytimes.com/2020/01/20/climate/climate-change-funding-states.html.

37. Or Should Not Be Named; in *Warmth,* Daniel Sherrell deals with the naming problem by labeling climate change only as "the Problem," and explores how using the more common phrase "can erase the thing itself . . . like trying to capture the whole of a book by invoking its title." Daniel Sherrell, *Warmth: Coming of Age at the End of Our World* (New York: Penguin, 2021), 175).

38. I have translated Krenz's words from the excerpts of his speech quoted in *Sachstand: Der Begriff "Wende" als Bezeichnung für den Untergang der DDR* (Berlin: Deutscher Bundestag, Wissenschaftliche Dienste, 2019), 3–4.

39. Krenz, in *Sachstand,* 4.

40. Jürgen-Friedrich Hake, Wolfgang Fischer, Sandra Venghaus, et al., "The German Energiewende—History and Status Quo," *Energy* 92 (2015): 532–546.

41. David Jacobs, "The German Energiewende: History, Targets, Policies and Challenges," *Renewable Energy Law and Policy Review* 3, no. 4 (2012): 223–233.

42. International Energy Agency, "Germany 2020: Energy Policy Review," executive summary, http://iea.org/reports/Germany-2020; Peter Friederici, "In Germany, the Energy Transition Continues," *Bulletin of the Atomic Scientists* 77, no. 2 (2021): 82–85.

CHAPTER 3: NARRATIVE

1. Daniel Sherrell, *Warmth: Coming of Age at the End of Our World* (New York: Penguin, 2021), 85.

2. Joan Didion, *The White Album*, in *We Tell Ourselves Stories in Order to Live: Collected Nonfiction* (New York: Alfred A. Knopf, 2006), 185.

3. Didion, *The White Album*, 185.

4. Roy Scranton, *Learning to Die in the Anthropocene: Reflections on the End of a Civilization* (San Francisco, CA: City Lights, 2015), 22.

5. Roy Scranton, *We're Doomed. Now What?* (New York: Soho Press, 2018), 5.

6. Frank Kermode, *The Sense of an Ending: Studies in the Theory of Fiction* (New York: Oxford University Press, 1967), 17.

7. Kermode, *The Sense of an Ending*, 46.

8. Annika Arnold makes a related point about climate change in her book *Climate Change and Storytelling: Narratives and Cultural Meaning in Environmental Communication* (Cham, Switzerland: Palgrave Macmillan, 2018), 124: "Climate change becomes part of the social world by mimicking those pre-modern structured myths and narratives our societies are still holding onto."

9. Amitav Ghosh, *The Great Derangement: Climate Change and the Unthinkable* (Chicago: University of Chicago Press, 2016), 80.

10. Ursula K. Le Guin, "The Carrier-Bag Theory of Fiction," in *Dancing at the Edge of the World* (New York: Grove Press, 1989), 165–166.

11. Jutta Bolt and Jan Luiten van Zanden, Maddison Project Database, 2020, https://www.rug.nl/ggdc/historicaldevelopment/maddison/releases/maddison-project-database-2020. Even graphed out per capita, the GDP is not good way to map economic progress on any finer scale than that of the entire nation-state, eliding as it does any breakdowns according to income level. Nor, as many critics have pointed out, is it a sensible way to measure national progress at all, ignoring as it does such metrics as happiness, health, and environmental sustainability.

12. Bolt and van Zanden, Maddison Project; Ralph F. Keeling, S. C. Piper, R. B. Bacastow, et al., "Atmospheric CO_2 Data," Scripps CO_2 Program, Scripps Institution of Oceanography, https://scrippsco2.ucsd.edu/data/atmospheric_co2/primary_mlo_co2_record.html.

13. Hartmut Rosa, *Social Acceleration: A New Theory of Modernity*, trans. Jonathan Trejo-Mathys (New York: Columbia University Press, 2013), 21.

14. C. Roser-Renouf et al., *Global Warming, God, and the "End Times"* (New Haven, CT: Yale Program on Climate Change Communication, 2016).

15. Robin Globus Veldman, *The Gospel of Climate Skepticism: Why Evangelical Christians Oppose Action on Climate Change* (Berkeley: University of California Press, 2019).

16. Jennifer O'Connell, "Why Silicon Valley Billionaires Are Prepping for the Apocalypse in New Zealand," *Guardian*, February 15, 2018; Bradley Garrett, *Bunker: Building for the End Times* (New York: Scribner, 2020).

17. Vivian Salama, Rebecca Ballhaus, Andrew Restuccia, et al., "President Trump Eyes a New Real-Estate Purchase," *Wall Street Journal*, August 16, 2019.

18. Christian Parenti, *Tropic of Chaos: Climate Change and the New Geography of Violence* (New York: Nation Books, 2011). Parenti is not original in reaching for the lifeboat metaphor. The ecologist Garrett Hardin, renowned for his deeply flawed but highly influential 1968 essay "The Tragedy of the Commons," was a white supremacist who contributed an essay titled "Lifeboat Ethics: The Case against Helping the Poor" to *Psychology Today* in 1974. It posited that wealthy countries needed to adopt harsh "lifeboat ethics"—that is, policies to keep others from climbing into the boat—in order to save themselves from being swamped by arrivals from poor countries.

19. Ghosh, *The Great Derangement,* 145.

20. Parenti, *Tropic of Chaos,* 11.

21. Ron Suskind, "Faith, Certainty and the Presidency of George W. Bush," *New York Times,* October 17, 2004, https://www.nytimes.com/2004/10/17/magazine/faith-certainty-and-the-presidency-of-george-w-bush.html. Neither Rove nor Suskind has ever confirmed that this quotation is actually Rove's. In the original magazine story the quotation is attributed to a White House "senior advisor." By 2007 the quotation was said to be "widely known" to be from Rove. Mark Danner, "Words in a Time of War: On Rhetoric, Truth and Power," in *What Orwell Didn't Know: Propaganda and the New Face of American Politics*, ed. Andras Szanto (New York: Public Affairs, 2007).

22. Matt Daily, "Exxon CEO Calls Climate Change Engineering Problem," Reuters, June 27, 2012, https://www.reuters.com/article/us-exxon-climate-idUSBRE85Q1C820120627.

23. Specifically, viewers who watch more fictional narratives evince a greater belief in a "just world," meaning that they believe that stories in the real world tend to work out OK. Markus Appel, "Fictional Narratives Cultivate Just-World Beliefs," *Journal of Communication* 58, no. 1 (March 2008): 62–83.

24. Brian Petersen, Diana Stuart, and Ryan Gunderson, "Reconceptualizing Climate Change Denial: Ideological Denialism Misdiagnoses Climate Change and Limits Effective Action," *Human Ecology Review* 25, no. 2 (2019): 117–141.

25. Ghosh, *The Great Derangement,* 30.

26. Kari Marie Norgaard, *Living in Denial: Climate Change, Emotions, and Everyday Life* (Cambridge, MA: MIT Press, 2011), xix, 207.

27. Frank Kermode explores this connection in *The Sense of an Ending,* 46–52.

28. Guy Debord, *The Society of the Spectacle*, trans. Donald Nicholson-Smith (New York: Zone Books, 1994), 17.

29. Debord, *The Society of the Spectacle*, 4, 10.

30. Robert Jay Lifton, *The Climate Swerve* (New York: New Press, 2017), 89–90.

31. Diana Stuart, Brian Petersen, and Ryan Gunderson, "Shared Pretenses for Collective Inaction: The Economic Growth Imperative, COVID-19, and Climate Change," *Globalizations* (2021), doi: 10.1080/14747731.2021.1943897.

32. Erin James, *The Storyworld Accord: Econarratology and Postcolonial Narratives* (Lincoln: University of Nebraska Press, 2015), xii.

33. James, *The Storyworld Accord*, 22.

34. Paul Slovic, "Perception of Risk," *Science* 236, no. 4799 (April 17, 1987): 280–285.

35. Emily Raboteau, "Climate Signs," *New York Review of Books*, February 1, 2019, https://www.nybooks.com/daily/2019/02/01/climate-signs.

36. Matthew Feinberg and Robb Willer, "Apocalypse Soon? Dire Messages Reduce Belief in Global Warming by Contradicting Just-World Beliefs," *Psychological Science* 22, no. 1 (2011): 34–38.

37. Panu Pihkala, "Eco-Anxiety, Tragedy, and Hope: Psychological and Spiritual Dimensions of Climate Change," *Zygon* 53, no. 2 (June 2018): 552.

38. Jamie Arndt, Sheldon Solomon, Tim Kasser, et al., "The Urge to Splurge: A Terror Management Account of Materialism and Consumer Behavior," *Journal of Consumer Psychology* 14, no. 3 (2004): 198–212.

39. Daniel Västfjäll et al., "Pseudoinefficacy and the Arithmetic of Compassion," in *Numbers and Nerves: Information, Emotion, and Meaning in a World of Data*, ed. Scott Slovic and Paul Slovic (Corvallis: Oregon State University Press, 2015).

40. Pihkala, "Eco-Anxiety, Tragedy, and Hope," 548, 549.

41. Per Espen Stoknes, *What We Think About When We Try Not to Think About Climate Change* (White River Junction, VT: Chelsea Green, 2015), 82.

42. Naomi Oreskes, Michael Oppenheimer, and Dale Jamieson, "Scientists Have Been Underestimating the Pace of Climate Change," *Observations* (blog), *Scientific American*, August 19, 201, https://blogs.scientificamerican.com/observations/scientists-have-been-underestimating-the-pace-of-climate-change.

43. David Spratt and Ian Dunlop, *What Lies Beneath: The Understatement of Existential Climate Risk* (Melbourne, Victoria: Breakthrough—National Centre for Climate Restoration, 2018).

44. Oreskes et al., "Scientists Have Been Underestimating the Pace of Climate Change."

45. Jonathan Mingle, "A World without Ice," *New York Review of Books*, May 14, 2020, https://www.nybooks.com/articles/2020/05/14/climate-change-world-without-ice.

46. Spratt and Dunlop, *What Lies Beneath*; Rowan T. Sutton, "Climate Science Needs to Take Risk Assessment Much More Seriously," *Bulletin of the American Meteorological*

Society 100, no. 9 (September 2019): 1637–1642; Bent Flyvbjerg, "The Law of Regression to the Tail: How to Survive Covid-19, the Climate Crisis, and Other Disasters," *Environmental Science and Policy* 114 (2020): 614–618.

47. Spratt and Dunlop, *What Lies Beneath*, 12.

48. Lesley Head and Theresa Harada, "Keeping the Heart a Long Way from the Brain: The Emotional Labour of Climate Scientists," *Emotion, Space and Society* 24 (August 2017): 34.

49. David Wallace-Wells, *The Uninhabitable Earth: Life after Warming* (New York: Tim Duggan Books, 2019).

50. Jem Bendell, "Deep Adaptation: A Map for Navigating Climate Tragedy," IFLAS Occasional Paper 2 (University of Cumbria, Institute for Leadership and Sustainability, July 2018, rev. July 2020), https://www.lifeworth.com/deepadaptation.pdf. For a critique, see Thomas Nicholas, Galen Hall, and Colleen Schmidt, "The Faulty Science, Doomism, and Flawed Conclusions of Deep Adaptation," Open Democracy, July 14, 2020, https://www.opendemocracy.net/en/oureconomy/faulty-science-doomism-and-flawed-conclusions-deep-adaptation.

51. Debord, *The Society of the Spectacle*, 127.

52. Debord, *The Society of the Spectacle*, 136.

53. Clive Hamilton, *Defiant Earth: The Fate of Humans in the Anthropocene* (Cambridge: Polity, 2017), 117–118.

54. Bruno Maçães, *History Has Begun: The Birth of a New America* (New York: Oxford University Press, 2020), 8, 105.

55. Elizier Yudkowsky, "Cognitive Bias Potentially Affecting Judgement of Global Risks," in *Global Catastrophic Risks*, ed. Nick Bostrom and Milan M. Ćirković (New York: Oxford University Press, 2008), 104.

56. Summarized in Joseph E. Uscinski, Stephan Lewandowsky, and Karen M. Douglas, "Climate Change Conspiracy Theories," in *The Oxford Research Encyclopedia of Climate Science*, Hans von Storch, editor in chief (New York: Oxford University Press, 2016).

57. Emily Nussbaum, "The Search for Pizazz at the Impeachment Reality Show," *New Yorker*, November 20, 2019, https://www.newyorker.com/culture/cultural-comment/the-search-for-pizzazz-at-the-impeachment-reality-show.

58. Kurt Andersen, *Fantasyland: How America Went Haywire: A 500-Year History* (New York: Random House, 2017), 9.

59. Aldo Leopold, *A Sand County Almanac, with Other Essays on Conservation from Round River* (New York: Oxford University Press, 1966), 183.

60. According to the Yale Program on Climate Change Communication, in 2020 just over a third of US poll respondents said they "often" or "occasionally" talked about global warming with friends or family, even as almost two-thirds said they are worried about it. Jennifer Marlon, Peter Howe, Matto Mildenberger, et al., "Yale Climate Opinion Maps 2020," Yale Program on Climate Change Communication, September 2, 2020, https://climatecommunication.yale.edu/visualizations-data/ycom-us/.

61. Elin Kelsey, *Hope Matters: Why Changing the Way We Think Is Critical to Solving the Environmental Crisis* (Vancouver, BC: Greystone Books, 2020), 61.

62. For example, Willcox A. Cunsolo, S. Harper, J. D. Ford, et al., "Ecological Grief and Anxiety: The Start of a Healthy Response to Climate Change?," *Lancet Planetary Health* 4 (July 2020), e261–e263; Panu Pihkala, "Anxiety and the Ecological Crisis: An Analysis of Eco-Anxiety and Climate Anxiety," *Sustainability* 12 (2020): 7836, doi:10.3390/su12197836.

63. Glenn A. Albrecht, *Earth Emotions: New Words for a New World* (Ithaca, NY: Cornell University Press, 2019), 38–39.

64. Joanna Macy and Sam Mowe, "The Work That Reconnects," Tricycle, Spring 2015, https://tricycle.org/magazine/work-reconnects.

65. For an overview, see Emma Lawrence, Rhiannon Thompson, Gianluca Fontana, et al., "The Impact of Climate Change on Mental Health and Emotional Wellbeing: Current Evidence and Implications for Policy and Practice," Grantham Institute Briefing Paper no. 36 (London: Grantham Institute, 2021), https://www.preventionweb.net/publications/view/78682.

66. See the website at https://www.goodgriefnetwork.org.

67. Ben Knight, "Coping with Climate Anxiety on a Warming Planet," *Deutsche Welle*, November 27, 2019, https://www.dw.com/en/coping-with-climate-anxiety-on-a-warming-planet/a-51198686.

CHAPTER 4: TRAGEDY

1. Paul O'Connor, "Nightfall in Atlantis," *Dark Mountain* 16 (2019): 11.

2. Joseph W. Meeker, *The Comedy of Survival: Studies in Literary Ecology* (New York: Charles Scribner's Sons, 1974), 41–42.

3. Meeker, *The Comedy of Survival*, 58.

4. Kate Raworth, *Doughnut Economics: 7 Ways to Think Like a 21st Century Economist* (White River Junction, VT: Chelsea Green, 2017).

5. Herman Daly, "Introduction to the Steady-State Economy," in *Economics, Ecology, Ethics: Essays toward a Steady-State Economy*, ed. Herman Daly (San Francisco, CA: W. H. Freeman & Co.), 5–6.

6. Raworth, *Doughnut Economics*, 13.

7. Quoted in Raworth, *Doughnut Economics*, 223.

8. Raworth, *Doughnut Economics*, 223–224.

9. Timothy Mitchell, *Carbon Democracy: Political Power in the Age of Oil* (London: Verso, 2013), 234.

10. William Appleman Williams, *The Great Evasion* (Chicago: Quadrangle Books, 1966), 12.

11. Shane Frederick, George Loewenstein, and Ted O'Donoghue, "Time Discounting and Time Preference: A Critical Review," *Journal of Economic Literature* 40, no. 2 (June 2002): 351–401.

12. Joseph E. Stiglitz, "Are We Overreacting on Climate Change?" *New York Times*, July 16, 2020, https://www.nytimes.com/2020/07/16/books/review/bjorn-lomborg-false-alarm-joseph-stiglitz.html.

13. Bruno Latour, *Down to Earth: Politics in the New Climatic Regime*, trans. Catherine Porter (Cambridge: Polity, 2018), 120.

14. Nicholas Stern, *The Stern Review: The Economics of Climate Change* (Cambridge: Cambridge University Press, 2007), viii. The original Stern Review on the Economics of Climate Change was commissioned by the UK government and released on October 30, 2006. Sir Nicholas Stern was formerly chief economist of the World Bank.

15. Stern, *The Stern Review*, 160.

16. Christopher Groves, "Living in Uncertainty: Anthropogenic Global Warming and the Limits of 'Risk Thinking,'" in *Future Ethics: Climate Change and Apocalyptic Imagination*, ed. Stefan Skrimshire (London: Continuum, 2010), 119.

17. Judith Nies, "The Black Mesa Syndrome: Indian Lands, Black Gold," *Orion* 17, no. 3 (Summer 1998): 18–29.

18. Frank Ramsey, "A Mathematical Theory of Saving," *Economic Journal* 38, no. 152 (December 1928): 543.

19. Some economists have begun reexamining Ramsey's ideas in light of climate change. Two examples are Francis Dennig, "Climate Change and the Re-evaluation of Cost-Benefit Analysis," *Climatic Change* 151 (2018): 43–54; and Partha Dasgupta, "Ramsey and Intergenerational Welfare Economics," *Stanford Encyclopedia of Philosophy*, June 1, 2019, https://plato.stanford.edu/entries/ramsey-economics.

20. International Monetary Fund, "Climate Change: Physical Risk and Equity Prices," in *Global Financial Stability Report: Markets in the Time of COVID-19*, April 2020, https://www.imf.org/en/Publications/GFSR/Issues/2020/04/14/global-financial-stability-report-april-2020.

21. Environmental humanist Erin James both discusses the limitations of the word *we* when it comes to climate change and suggests expanding the definition of the word *narrative* to include elements of the more-than-human world: "Narrative in the Anthropocene," in *Environment and Narrative: New Directions in Econarratology*, ed. Erin James and Eric Morel (Columbus: Ohio State University Press, 2020), 183–202.

22. Taxpayers for Common Sense, *Padding Big Oil's Profits*, February 2020, https://www.taxpayer.net/wp-content/uploads/2020/02/TCS-Padding-Big-Oil-Profits_Feb.-2020.pdf.

23. Robert J. Brulle, Melissa Aronczyk, and Jason Carmichael, "Corporate Promotion and Climate Change: An Analysis of Key Variables Affecting Advertising Spending by Major Oil Corporations, 1986–2015," *Climatic Change* 159 (2020): 87–101.

24. Geoff Dembicki, "The Multi-Billion-Dollar 'Climate Services' Industry," *Ensia*, August 2, 2019, https://ensia.com/features/private-climate-services-industry-environmental-justice-corporations-inequity.

25. Svenja Keele, "Consultants and the Business of Climate Services," *Climatic Change* 157 (2019): 12.

26. Naomi Klein, *The Shock Doctrine: The Rise of Disaster Capitalism* (New York: Henry Holt, 2007), 419.

27. Hannah Arendt, Letter to Gershom Gerhard Scholem, Hannah Arendt Papers at the Library of Congress, Correspondence—Scholem, Gershom Gerhard—1963–1964, n.d., Adolf Eichmann File, 1938–1968.

28. Clive Hamilton, *Defiant Earth: The Fate of Humans in the Anthropocene* (Cambridge: Polity, 2017), 151.

29. Greg Grandin, *The End of the Myth: From the Frontier to the Border Wall in the Mind of America* (New York: Metropolitan Books, 2019), 273. The "brazen show" language appears in Hiroko Tabuchi, "'Rolling Coal' in Diesel Trucks, to Rebel and Provoke," *New York Times*, September 4, 2016.

30. In his article "Quantifying National Responsibility for Climate Breakdown" Jason Hickel ascribes 40 percent of "excess global CO_2 emissions" since 1850 to the United States, with the EU nations responsible for another 29 percent.

31. Cara Daggett, "Petro-masculinity: Fossil Fuels and Authoritarian Desire," *Millennium: Journal of International Studies* 47, no. 1 (2018): 25–44.

32. Meeker, *The Comedy of Survival*, 36–37, 51.

33. Marek Oziewicz, "Fantasy for the Anthropocene: On the Ecocidal Unconscious, Planetarianism, and Imagination of Biocentric Futures," in *Fantasy and Myth in the Anthropocene: Imagining Futures and Dreaming Hope in Literature and Media*, ed. Marek Oziewicz, Brian Attebery and Tereza Dědinova (New York: Bloomsbury Academic, 2022), 58.

34. William Cronon, "A Place for Stories: Nature, History, and Narrative," *Journal of American History* 78, no. 4 (March 1992): 1350.

35. Nina Eliasoph, *Avoiding Politics: How Americans Produce Apathy in Everyday Life* (Cambridge: Cambridge University Press, 1998), 17.

36. Erich Fromm, *Escape from Freedom* (New York: Farrar & Rinehart, 1941).

37. International Energy Agency, Global CO_2 Emissions in 2019, https://www.iea.org/articles/global-co2-emissions-in-2019.

38. James D. Ward, Paul C, Sutton, Adrian D. Werner, et al., "Is Decoupling GDP Growth from Environmental Impact Possible?," *PLOS One* 11, no. 10 (2016), e0164733, doi: 10.1371/journal.pone.0164733.

39. Jason Hickel and Giorgos Kallis, "Is Green Growth Possible?," *New Political Economy* 25, no. 4 (2020): 469–486.

40. Jason Hickel, *Less Is More: How Degrowth Will Save the World* (London: William Heinemann, 2020), 22.

41. Diana Stuart, Ryan Gunderson, and Brian Petersen, *The Degrowth Alternative: A Path to Address our Environmental Crisis?* (Abingdon-on-Thames: Taylor and Francis, 2020).

42. Hickel, *Less Is More*, 201.

43. Giorgos Kallis, "The Degrowth Alternative," Great Transition Initiative, 2015, https://greattransition.org/publication/the-degrowth-alternative.

44. Hickel, *Less Is More*, 177–181.

45. Eric Holthaus, *The Future Earth: A Radical Vision for What's Possible in the Age of Warming* (New York: HarperOne, 2020), 128.

46. Hickel, *Less Is More*, 30–31.

CHAPTER 5: COMEDY AND COMPLEXITY

1. David Graeber and David Wengrow, *The Dawn of Everything: A New History of Humanity* (New York: Farrar, Straus and Giroux, 2021), 133.

2. Amitav Ghosh makes a parallel point, if with a different spatial metaphor, in *The Great Derangement: Climate Change and the Unthinkable* (Chicago: University of Chicago Press, 2016), examining how mainstream novelists in particular have largely ignored climate change, leaving it for other writers to examine: "It was in exactly the period in which human activity was changing the earth's atmosphere that the literary imagination became radically centered on the human. Inasmuch as the nonhuman was written about at all, it was not within the mansion of serious fiction but rather in the outhouses to which science fiction and fantasy had been banished" (66).

3. Joyce Carol Oates, "Against Nature," in *On Nature: Nature, Landscape, and Natural History*, ed. David Halpern (San Francisco, CA: North Point Press, 1987), 236.

4. Nicholas Georgescu-Roegen, *La Décroissance: Entropie, écologie, économie*, 3rd ed. (Paris: Sang de la Terre et Ellébore, 2006), quoted in Pablo Servigne and Raphaël Stevens, *How Everything Can Collapse: A Manual for Our Times*, trans. Andrew Brown (Cambridge: Polity, 2020), 181.

5. Christy Rodgers explores similar terrain in "At Play in the Comedy of Survival: An Appreciation of Joseph Meeker," Dark Mountain Project, 2015, https://dark-mountain.net/at-play-in-the-comedy-of-survival-an-appreciation-of-joseph-meeker.

6. Joseph W. Meeker, *The Comedy of Survival: Studies in Literary Ecology* (New York: Charles Scribner's Sons, 1974), 22.

7. Meeker, *The Comedy of Survival*, 23, 24, 26.

8. Meeker, *The Comedy of Survival*, 27.

9. Randy Olson, *Don't Be Such a Scientist: Talking Substance in an Age of Style* (Covelo, CA: Island Press, 2009), 32–39.

10. Brian K. Hall, "The Paradoxical Platypus," *BioScience* 49, no. 3 (March 1999): 211–218.

11. Paula Spaeth Anich, Sharon Anthony, Michaela Carlson, et al., "Biofluorescence in the Platypus," *Mammalia* (2020), doi: https://doi.org/10.1515/mammalia-2020-0027.

12. Max Boykoff and Beth Osnes, "A Laughing Matter? Confronting Climate Change through Humor," *Political Geography* 68 (2019): 154, 155.

13. P. R. Brewer and J. McKnight, "Climate as Comedy: The Effects of Satirical Television News on Climate Change Perceptions," *Science Communication* 37, no. 5 (2015): 635–657.

14. Nicole Seymour, *Bad Environmentalism: Irony and Irreverence in the Ecological Age* (Minneapolis: University of Minnesota Press, 2018), 234.

15. Meeker, *The Comedy of Survival*, 39.

16. Jan Zwicky, "A Ship from Delos," in Robert Bringhurst and Jan Zwicky, *Learning to Die: Wisdom in the Age of Climate Crisis* (Regina, Saskatchewan: University of Regina Press, 2018), 70.

17. Victor Turner, *From Ritual to Theatre: The Human Seriousness of Play* (New York: PAJ Publications, 1982).

18. Celia Deane-Drummond, "Beyond Humanity's End: An Exploration of a Dramatic Versus Narrative Rhetoric and Its Ethical Implications," in *Future Ethics: Climate Change and Apocalyptic Imagination*, ed. Stefan Skrimshire (London: Continuum, 2010), 246, 252.

19. Deane-Drummond, "Beyond Humanity's End," 253.

20. Rob Nixon identifies this tendency in *Slow Violence*, pointing to examples of picaresque and bricolage literary works created in opposition to exploitative corporate power in the developing world.

21. Mary Annaïse Heglar, "Climate Change Isn't the First Existential Threat," Zora, February 18, 2019, https://zora.medium.com/sorry-yall-but-climate-change-ain-t-the-first-existential-threat-b3c999267aa0.

22. Graeber and Wengrow, *The Dawn of Everything*.

23. Donna Haraway, *Staying with the Trouble: Making Kin in the Chthulucene* (Durham, NC: Duke University Press, 2016), 86.

24. Jonathan Lear, *Radical Hope: Ethics in the Face of Cultural Devastation* (Cambridge, MA: Harvard University Press, 2006), 78.

25. Lear, *Radical Hope*, 83.

26. Lear, *Radical Hope*, 95.

27. Kathleen Dean Moore, *Great Tide Rising: Towards Clarity and Moral Courage in a Time of Planetary Change* (Berkeley, CA: Counterpoint, 2016), 213.

28. Moore, *Great Tide Rising*, 305.

29. Aja Romano, "Hopepunk, the Latest Storytelling Trend, Is All About Weaponized Optimism," Vox, December 27, 2018, https://www.vox.com/2018/12/27/18137571/what-is-hopepunk-noblebright-grimdark.

30. Quoted in John Foster, *Towards Deep Hope: Climate Tragedy, Realism and Policy* (Weymouth, Dorset, UK: Green House, 2017), 17.

31. Julia B. Corbett, *Communicating the Climate Crisis: New Directions for Facing What Lies Ahead* (Lanham, MD: Lexington Books, 2021), 192.

32. James Button, "The Climate Interviews," *The Monthly*, March 2020, 34.

33. Henry David Thoreau, *The Maine Woods* (Boston: Ticknor and Fields, 1864), 93.

34. Emily Atkin, *Heated,* April 1, 2020, https://heated.world/p/episode-1-bill-mckibben-on-solidarity.

35. William R. Jordan III, *The Sunflower Forest: Ecological Restoration and the New Communion with Nature* (Berkeley: University of California Press, 2003), 183.

36. Ursula K. Le Guin, "The Carrier-Bag Theory of Fiction," in *Dancing at the Edge of the World* (New York: Grove Press, 1989), 169, 168.

37. Robert D. Putnam and Shaylyn Romney Garrett, *The Upswing: How America Came Together a Century Ago and How We Can Do It Again* (New York: Simon & Schuster, 2020), 317.

38. Haraway, *Staying with the Trouble*, 4.

39. Kendra Pierre-Louis, "Wakanda Doesn't Have Suburbs," in *All We Can Save: Truth, Courage, and Solutions for the Climate Crisis*, ed. Ayana Elizabeth Johnson and Katharine K. Wilkinson (New York: One World, 2020), 141.

40. Pierre-Louis, "Wakanda Doesn't Have Suburbs," 144.

41. Bruno Latour, *Down to Earth: Politics in the New Climatic Regime*, trans. Catherine Porter (Cambridge: Polity, 2018), 12.

42. Ursula K. Heise, *Sense of Place and Sense of Planet: The Environmental Imagination of the Global* (New York: Oxford University Press, 2008), 61.

43. Jonathan Mingle, "A World without Ice," *New York Review of Books*, May 14, 2020, https://www.nybooks.com/articles/2020/05/14/climate-change-world-without-ice.

44. Emily Johnston, "Loving a Vanishing World," in Johnson and Wilkinson, ed., *All We Can Save*, 258.

45. Quoted in Glenn A. Albrecht, *Earth Emotions: New Words for a New World* (Ithaca, NY: Cornell University Press, 2019), 53.

46. Rebecca Solnit, *A Paradise Built in Hell: The Extraordinary Communities That Arise in Disaster* (New York: Penguin, 2010).

47. Sarah Jaquette Ray, *A Field Guide to Climate Anxiety* (Berkeley: University of California Press, 2020), 7.

48. Ilona M. Otto, Jonathan F. Donges, Roger Cremades, et al., "Social Tipping Dynamics for Stabilizing Earth's Climate by 2050," *PNAS* 117, no. 5 (February 4, 2020): 2354–2365.

INDEX